Patterns of Light

A nearly complete overview of this book can be found in Archimedes' Burning-Mirrors from Witelo's *Optics* written c. 1270. In the woodcut, we see light traveling in straight lines, reflection, mirrors, images, bent legs in water, rainbows, blue sky, and clouds. The rest of the electromagnetic spectrum, invisible to the eye, yields no fewer clues about both microscopic and distant worlds. The optical properties of elephants we leave to the valiant.

Steven Beeson and James W. Mayer

Patterns of Light

Chasing the Spectrum from Aristotle to LEDs

 Springer

Steven Beeson
Waterfront Technologies
Washington, DC 20010
USA
sbeesonsf@yahoo.com

James W. Mayer
Center for Solid State Science
Arizon State Univeristy
Tempe, AZ 85287-1704
USA
james.mayer@asu.edu

ISBN 978-0-387-75106-1   e-ISBN 978-0-387-75107-8

Library of Congress Control Number: 2007936185

Printed on acid-free paper.

9 8 7 6 5 4 3 2 1

springer.com

"Come forth into the light of things. Let nature be your teacher."

-- Wordsworth, 1788

"The electron is not as simple as it looks."

--W. Lawrence Bragg

Steve Beeson:

To my parents, Donald and Luana, who always see the light inside

Jim Mayer:

To my wife, Elizabeth

Preface

Light is all around us. Vision is our dominant sense, and we are richly rewarded with a palette of colors from red to violet. Our eyes do not detect the low-energy, long-wavelength infrared (IR) radiation, but we know it exists from discussions of war applications and televised images of guided weapons targets. Our eyes do not detect the higher-energy (above visible light energies) and shorter-than-visible-wavelength ultraviolet radiation, and yet we know it is there from the sunburn we receive in Arizona. We also know that window glass can block ultraviolet rays so we don't get a burn while driving with the windows rolled up.

We know about radio waves from the little boxes that talk to us and x-rays from the dentist office. These waves and rays belong to the same family of light, often called photons (from the Greek *photos*, light), that describes the spectra of electromagnetic radiation over 10 orders of magnitude from very low-energy radio waves to very high-energy x-rays and gamma rays.

This book starts with the visible — the straight path of light — because what we can see is a good starting point. It continues with reflection as we look at ourselves in mirrors and storefront windows. It includes a chapter on refraction, its discovery and description, and a chapter on lenses, which are familiar to the myopic (eyeglasses) and the football fan (binoculars). Color is introduced with the query, Why is the sky blue? After answering that and other similar questions, the book goes beyond the visible to the infrared and ultraviolet. It ends with analysis of Mars using x-ray emission.

This is a descriptive book rather than a technical book. It is designed for the general reader with no background in science but who has an interest in the light around us. There is an Internet site associated with the text called Images of Nature (http://ion.eas.asu.edu). It provides optical and electron microscopy images that can be downloaded by the reader and student for free. The site also provides additional information on x-ray emission techniques and comparisons between optical and electron microscopy. For the educator or curious reader, exercises, solutions, and exams are available by writing to the publisher.

The book originated in a general studies course at Arizona State University in the curriculum of Physical Science and in that of Society, Values and Technology. The course started with the support of Terry Woodin of the National Science Foundation for K–12 teachers. It expanded to the general undergraduate population at Arizona State University and its enrollment is at 150 students each semester (enrollment started with 10 students in Elementary Education). Verne Hess of NSF supported Images.

At Arizona State University we want to thank Sue Wyckoff for her support through the Arizona Collaborative for Excellence in Teacher Preparation. We thank Elizabeth Mayer, who participates in the ASU course; Frank Mayer for handling the lab sessions; Rod Heyd, our web-master; and Misty Wing for manuscript preparation. Michael Weiland and Marc van Horne provided IT course support.

Thanks to Danny Adams and Rudolph Nchodu of the University of the Western Cape and Frank Mayer for the three years of Patterns workshops given to high school teachers in South Africa. Images of Nature arose out of their interest.

Steve Beeson would like to acknowledge the US Department of Labor for the time and computing resources in finishing this work. Additional thanks go to Alison, Lisa, and Lilly for advice; to Tyler for use of the 'vette; and to Laura and Karen for inspiration.

The line drawings were done by Jane Jorgensen and Ali Avcisoy. They were redrawn by Don Thompson of Graphixx.

T. von Forster read the original version and made extensive suggestions, which lead to the final manuscript.

Linda Young edited the manuscript. She has a sharp eye and made many corrections and improvements to the present manuscript.

James W. Mayer   Steve Beeson
Tempe   Washington, DC
May 2007   May 2007

Contents

1 The Path of Light

1.1 The Straight and Narrow

What do you notice about the photograph below? Beautiful beams of sunlight stream into New York's Grand Central Station from ornate windows high overhead. People gather in groups talking, discussing the day's events as they prepare to diverge from this place, to travel across the state or the country. From an astute observer's standpoint, these sunbeams stream in with absolutely straight

Fig. 1.1 Sunbeams illuminate travelers in Grand Central Station in New York. (Photo courtesy of Underwood Archives)

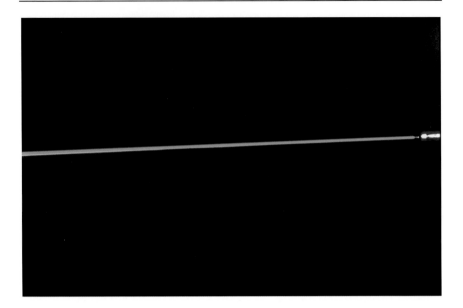

Fig. 1.2 A laser beam pierces a dust-filled room. Light can reflect from small particles such as dust, smoke, or haze to make the light beam visible. (Photo by J. Phillips)

paths. A question comes to mind: Why does the light travel straight?

The answer requires speculation and some philosophy. Light travels in a perfectly straight line simply because it is the most economical and efficient way for it to travel. Granted, this explanation leads to further questions, but save invoking quantum physics, Pierre de Fermat's Principle of Least Time explains the path of light through a medium and at an interface between two media.

A new question: How do we know light travels in a straight line? Since a beam of light itself is transparent to us unless it's actually shining in our eyes, we need to see light as it travels through a medium that will reflect some of the light back to us. Can you see sunlight in the night sky as it travels from the sun to the moon? No. It's traveling through empty space. We can't see it until it has something to reflect from, such as the Moon. Figure 1.2 shows a beam from a green laser traveling through a dust-filled room. The dust acts like small reflectors that send some of the light back to our eyes, allowing us to see the beam. Smoke also works well; many theater productions, films, and television shows use stage smoke with flashlights to dramatically delineate a light beam. What would detective or mystery shows be without the dusty room and bright flashlight?

The presence of dust in the atmosphere contributes to beautiful desert sunsets. In Fig. 1.3, dust and haze in the air reflect the sunlight peeking through the

Fig. 1.3 Clouds, haze and sunlight contribute to the stunning effect of light rays over the ocean. The dark areas between the beams — shadows caused by clouds — exhibit the straight path that light takes through the atmosphere. (Photo by E. Mayer)

clouds and into the observer's eye. It is remarkable that even over such long distances as the clouds to land, we can distinguish this fascinating and mysterious property of light.

Whether it is traveling through air, water, glass, diamond, a smoky Broadway stage, or any other transparent substance (or in nothing — the vacuum of space), light travels in a straight path until it encounters a different medium. So straight that analogies fail — the path of light is the Ultimate Straight Line.

1.2 The Fastest Thing Around

Besides its perfect path, light also has the ultimate cruise control. It travels straight at a constant speed called, appropriately enough, the *speed of light*, which in a vacuum is 3×10^8 meters per second (or 186,000 miles per second). Its velocity is a bit slower in transparent materials such as water or glass. Water slows light down by just a few thousand meters per second (or a few hundred miles per second). Diamond slows the light down by half, causing it to bend tremendously at the air/diamond interface. A more detailed discussion of diamond and its properties relating to light are found in Chapter 4.

Why 186,000 miles per second? Why not 37 miles per second? Or 10 billion miles per second? Or infinitely fast? This, too, is a question tough to answer. The speed of light is one of those properties of nature that seem to be immutable, fixed in the pantheon of *fundamental constants*, those most basic numbers which serve as the control rods for all physical and chemical processes. Recent discoveries, however, have shown that some of the supposedly fixed fundamental constants may have evolved (and might still be evolving, albeit extremely slowly) throughout the lifetime of the universe. Could the speed of light have been different in the past? If so, the implications would be staggering, forcing us to rethink not only the age and size of the universe, but the most basic principles of all of physics.

Regardless of its possible past value(s), the speed of light remains the upper limit of speed that all objects can attain. Nothing can travel faster than light[1]. In comparison, sound through air — the tortoise to light's hare — travels a paltry one-fifth mile per second. On a clear day, watch a carpenter pound a nail into the frame of a new building. What do you perceive first, the sight of the hammer hitting the nail, or the *sound* of the hammer hitting the nail? Hopefully, by a long shot, you see it first. The Ultimate Speed is 3×10^8 meters per second.

If light travels so fast, how did anyone measure its speed? How do we really know what its velocity is? I mean, you can't exactly time light doing laps on a quarter-mile track, can you? Actually a few people did, in a sense, and they didn't even have expensive equipment! All you need is either really great distances (larger-than-the-Earth distances) or really accurate timepieces.

In the seventeenth century, determining longitude — the distance east or west from an arbitrary Great Circle around the Earth — remained a vexing problem, especially for sailors. (*Latitude* is easy to determine: measure the height of the noontime sun above the horizon on the days of the equinoxes.) The proper measurement of longitude, however, remained elusive because it required accurate clocks at two locations.

Ole Roemer, a Danish astronomer, wanted to use the light of Jupiter's moons as they disappeared and reappeared behind the giant planet to determine how far the Earth had spun in a given time (and thus measure one's change in longitude in the same time). Instead, he ended up measuring the speed of light as it traveled across the solar system, and he used only a telescope and a clock! He observed the moons of Jupiter as they emerged from the far side of the gaseous planet, measuring their *periods*, or the time it takes to make one revolution. He noticed that if the Earth was on one side of its orbit, traveling toward Jupiter, the period of a certain moon would be eleven minutes faster than average. And if the Earth was on the other side of its orbit (six months later), traveling away from Jupiter, the period of the Jovian moon was eleven minutes slower. Roemer knew that, if light traveled infinitely fast, there should be no discrepancy

---

[1] Nothing, that is, that carries information, as light does.

between the sightings of the moon on either side of the Earth's orbit. Parting from popular opinion, Roemer was the first to show that light had a finite (but truly great) speed.

Three centuries later, Albert Michelson used a sophisticated timepiece to make the first modern and truly accurate measurement of the speed of light. Using a spinning mirror as a rudimentary clock, he pinpointed the time it took for light to travel between two mountaintops north of Los Angeles. Knowing the distance between the two mountaintops (22 miles), he easily determined the light's velocity. Michelson again experimented with light when he used another spinning apparatus to single-handedly discard the idea that light waves need to travel through a medium, the *luminiferous ether*, in order to propagate. We will leave more detailed discussions to later chapters.

1.3 Standing in the Shadows

Light would not be as interesting if it remained unsullied. Shadows exhibit the three-dimensional world to us through depth, distance, and contrast. Twelve hours of our busy day is spent in the shadow of the Earth. Shadows are, simply

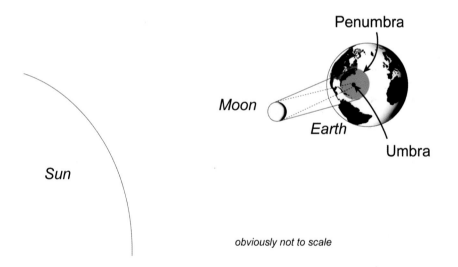

obviously not to scale

Fig. 1.4 An extended light source, like the Sun, casts many shadows. The sum of all the shadows of an object is the umbra. The outer edge, where only some of the shadows overlap, is the penumbra.

enough, areas where light from a source is blocked. A perfectly sharp shadow can be achieved with a perfect *point source* of light, a rare beast indeed. (If stars were bright enough in our light-polluted sky they would make wonderful shadow-casters.) Most light however, comes from *extended* sources: a light bulb, a candle, the Sun. Think of these extended sources as being made of many point sources, each point source casting its own shadow of the object. The sum of all the different shadows from the extended source is the shadow of the object, usually dark in the center, and fuzzy or dimmer on the edges.

Notice the fainter shadow around the darker shadow of your hand near a light bulb: The area on the edges of the shadow, where some of the shadows overlap, is called the *penumbra*. The darkest area, where all the shadows overlap, is called the *umbra*. Nature's most vivid display of shadows, umbra, and penumbra occur with *solar eclipses*. Figure 1.4 illustrates how the light from the Sun is

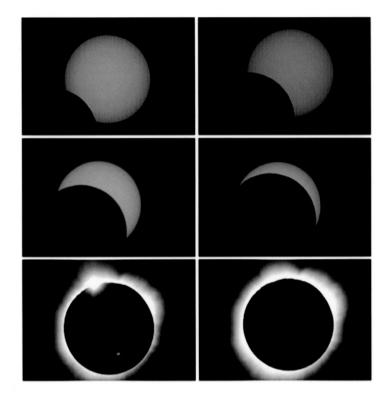

Fig. 1.5 Phases of the total solar eclipse of June 2001 as seen from Zimbabwe. (Photo courtesy of Reuters/Howard Burditt/Hulton/Archive)

blocked by the Moon, which casts a shadow on the sunward side of the Earth. Since the Sun is an extended source, viewers in the penumbra see a partial eclipse; those completely in the umbra see a total eclipse. A view from the umbra of the solar eclipse of June 2001 in Zimbabwe is shown in Fig. 1.5. Eclipses have been observed, feared, and studied throughout much of human history, and astronomers have understood the eclipse cycle for more than four hundred years. Currently there are many software packages and online resources that list eclipse data for at least the next four hundred years.

1.4 The Reversible Path of Light

In a material with uniform composition, such as glass, the path of light is straight, and the velocity of light is constant. When light enters another material, such as air or water, both its path and velocity change. The changes occurs at the boundary, or *interface*, between the two materials, similar to the way a toy wagon, rolling at an oblique angle from the sidewalk to the grass, will change its path and speed at the grassy edge. The path of light can undergo many changes as it travels from its source to the eye. To analyze its path, it is useful to know

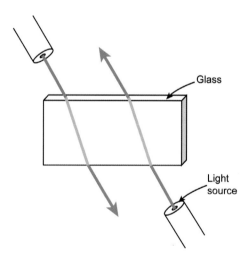

Fig. 1.6 Two light beams traveling through a rectangular block of glass. The beams are traveling in opposite directions, but they enter and exit at the same angle. If one light beam was aimed along the same path as the other exiting beam, they would be indistinguishable — the light paths are reversible.

that the path is reversible. That is, if you trace the path from water to air to glass, you can reverse the direction and follow the same path from glass to air to water. Figure 1.6 shows two paths of light traveling through a block of glass in opposite directions. Scientists use the reversibility concept to help test their hypotheses about the properties and behavior of light.

1.5 The World through a Hole

A simple demonstration of the path of light requires the construction of a very simplified camera. This camera has no lenses, requires no film, and costs only pennies to make. It works, however, on the same principles as very expensive cameras. All you need are two small cardboard tubes, some regular (non-clear) tape (such as Scotch^{TM} brand), a square of aluminum foil, and a pin.

To construct the pinhole camera, tape aluminum foil across one end of what will be the outer cardboard tube, and put a small pinhole in the foil. On the inner tube, which will slide in and out of the outer tube, secure the clear tape so that it covers one end of the tube. Place the inner tube inside the outer tube so that the inner tube's tape is closest to the outer tube's pinhole. The inner tube is called

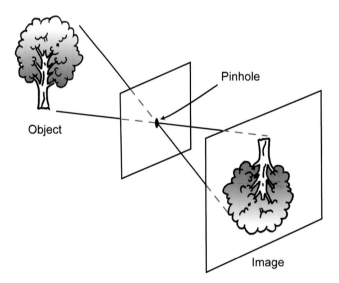

Fig. 1.7 A pinhole casts an image of a tree on a piece of paper. The straight lightbeams cross at the pinhole, causing the image to appear upside down.

the *viewer*. As you look through the viewer, you should see a bright light source on the tape. How does this happen? How could the light from the source get through the tiny pinhole to create a complete image of the source — especially when the pinhole is so much smaller than the object?

All visible objects either emit or reflect light. Some do both, but most just reflect. We see things by the light that is reflected off of them and into our eyes. When an object either emits or reflects light, the light travels in all directions from the object. You can confirm this by just walking around this book and looking at it; you can see it from every angle. However, only a tiny fraction of the light from the object gets into our eyes at one time. Even then, we are usually focusing on one part of the object and not the whole thing at once (unless it's far away).

Imagine a tree in the distance. We may focus on the top branches of the tree, but light from the trunk and the branches facing us still will travel into our eyes. The same thing happens with the pinhole and the light source. So why is the image upside down on the screen? Here is where we must think about the path of light. Aim your pinhole viewer at a tree and notice that the tree's image on the tape screen is upside down. Imagine a ray of light from the top branch of the tree traveling toward the pinhole. Since the pinhole is so small, only this ray and a few others from the top branch can pass through the hole. The light rays haven't

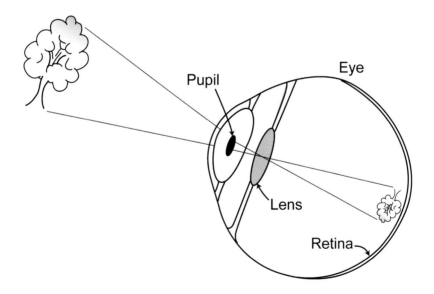

Fig. 1.8 The eye acts like a pinhole camera. The pupil is the pinhole, and the retina is the screen. The images projected on the retina are inverted.

encountered any different media, so they travel in a straight path from the tree branch to the screen. Notice in Fig. 1.7 that the rays from the top branch and from the trunk have to cross at the pinhole in order to pass through; the crossed light rays cause the image of the tree (or light bulb, or any other object) to be upside down. The pinhole is acting as a lens.

Your eyes work in very similar fashion to the pinhole camera. Figure 1.8 illustrates how the *pupil* of an eye is the primary light gatherer in the system. The eye's *crystalline lens* is used to fine-tune the focus of the image. And like the tape screen of our pinhole camera, the *retina* captures an upside-down image of an object. It's up to the brain to interpret the signal and right the image. We will discuss the eye in greater detail in Chapters 5 and 6.

Let's enlarge the pinhole in our homemade camera. Does the image become clearer or less clear? Does it get brighter or dimmer? Now punch a new, larger hole in the pinhole viewer and observe the light source through the viewer. You'll notice that the image becomes brighter but less clear. Imagine making the pinhole larger and larger while the image becomes less and less clear. Eventually, the hole is as large as the tube and the image is completely out of focus: the light rays do not cross any longer, and no image is formed. Images cannot form without light rays crossing.

1.6 A Room with a View

When you walk into the small, camera-shaped building a few steps from the Musée Mechanique near San Francisco's Cliff House, your eyes struggle to adjust to the darkness. A large parabolic disk rests lazily in the middle of the room, somehow connected to the concept inherent in the building's name, "Giant Camera." After a few moments in the dark, you understand what the disk reveals: a near perfect, live representation of the ocean landscape outside the room's walls. Through the hole in the ceiling, a small periscope projects onto the disk an image of whatever it's pointed to: Seal Rock, tourists strolling by the Cliff House, or Ocean Beach below. You are inside a *camera obscura*, a room-sized pinhole camera.

The simplest camera obscuras are not much more than the pinhole camera you made. If you've ever used a shoebox with a hole in one side to view a solar eclipse, you've been inside a simple camera obscura. Like the pinhole camera, a camera obscura uses the concept of light rays traveling in straight lines and crossing at holes to create an image of the outside scene. Unless a lens is used to re-invert the image, the scene projected is upside down, also like the pinhole camera. The lens also supplies more light to the image by allowing the *aperture* (or opening) to be larger than a pinhole would allow. The image from the camera obscura is dim but in sharp focus, a true "motion picture" more accurate than the view through the lens of a modern camera. The modern camera is, in fact, nothing more than a fancy camera obscura, from which it received its name.

Invented possibly as early as the twelfth century (or earlier, by the Chinese), the camera obscura is known to have been used extensively in the sixteenth and seventeenth centuries by artists and geographers interested in capturing the subtleties of light in landscapes or interior scenes. Several historians have argued that a few of the Dutch masters, including Johannes Vermeer and Jan van Eyck, used camera obscuras much like the one in Fig. 1.9 to render their masterpieces of everyday seventeenth-century European life: Vermeer's *Girl with a Red Hat* and *View of Delft*, and van Eyck's *Betrothal of Arnolfini* support this hypothesis. Regarding Vermeer's use of light in *Girl with a Pearl Earring*, critic Lawrence Gowing writes (1952), "The radiography of painting has indeed never shown a form in itself as wonderful as this strange, impersonal shape. We are in the presence of the real world of light, recording, as it seems, its own objective print."

In the comics of the 1930s and '40s, Superman used his x-ray vision to see through walls and clothing, finding the hidden gun and the bad guy's hideout. Behind the comic book flair lurks the historic and oft-repeated misconception that we see because it travels *from* our eyes *to* an object. If this were true, beams

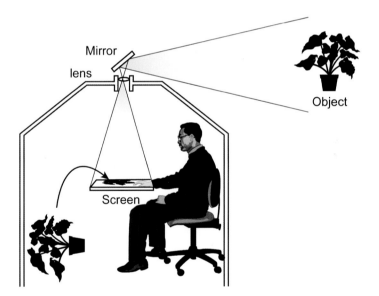

Fig. 1.9 A diagram of a camera obscura using a tent, a small opening in the ceiling, and a tube containing a lens and a mirror to project the scene outside. Art historians believe many Renaissance artists used camera obscuras to copy, highlight, or even trace their subjects. (Steadman 2001)

of light would emanate from our eyes, and the night sky would be filled with billions of searchlights. We'd blind our lover when we looked into his or her eyes. But that's not what happens. Instead, nature has built the eyeball in such a way as to take light in and process it. Not only does the eye soak up light, it uses the basic properties of light to help us better perceive our world: straight light beams cross at the pinhole-like pupil, shadows provide visual and emotional depth, and the great speed of light connects us virtually instantly to events. Next we'll see how the properties of light reflection have allowed us to take the eye one step further and recreate visual images with mirrors.

2 The Reflection of Light

2.1 Reflections on the Past

Besides the occasional light bulb, electric eel or star, we see things because light is reflected from them, a statement profound in its simplicity. Imagine a sneaker. If there is no light, is the sneaker there? Of course it is. We can touch it, hear it if it drops, sometimes smell it. But it casts no light of its own; it needs a light source to illuminate it. The light is then reflected[1] from the sneaker into our eyes, and the sneaker information is transmitted into our brain: color (white with brown dirt spots and red markings); shape (oblong, flat-bottomed); texture (rough, synthetic). Like the proverbial falling tree in the forest, the sneaker needs a mechanism and a receiver in order to be perceived. The receiver is our eye, the mechanism is light. The tree (and the sneaker) just reflects the light.

The ancients did not know that objects simply reflect light. Three thousand years ago, Plato and Euclid believed light exited the eye and somehow illuminated the object like Superman's x-ray vision, thus allowing it to be perceived. Empedocles in the fifth century BCE theorized that objects emit rays of information, which our eyes received by sending out a similar "visual ray" to intercept the information, as in Fig. 2.1(b). Eyes were thought literally to shine. The combined visual ray and information were then thought to be somehow transmitted to the eye, and an image was created in the mind.

Others believed objects sent out ultra-thin, invisible images of themselves in all directions, and that those were then received by the eye. These ethereal object-ghosts were thought to be transmitted in a continuous manner so that we would perceive an unbroken image of the object. Figure 2.1(c) shows an image being emitted in all directions, with some of the images landing in the eye of the observer. Many of these theories lasted into the seventeenth century. Students today still have difficulty understanding the mechanism of light reflection and visual perception. Ask a fourth grader how we see objects and you will likely hear the Superman Theory.

---

[1] Some is absorbed.

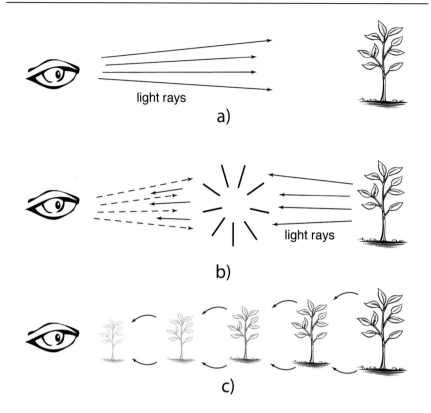

Fig. 2.1 An illustration of three theories of visual perception throughout history. **(a)** The Superman Theory, in which light is emitted from the eye, illuminating objects. **(b)** The object emits rays of information which the "visual rays" from the eye intercept and then transmit to the eye. **(c)** Ethereal images of an object are emitted in all directions — some land in the eye.

Not until Johannes Kepler in 1604 was the theory that light enters the eye simply as light treated in the proper scientific manner (i.e., based on an experimental process). Yet even after looking through the back of an eyeball and seeing an upside-down image, he still regarded light as a form of heat. After Kepler, most scientists agreed that light must only enter, not exit, the eye in order for an object to be visually perceived. Subsequently, it was the nature of light, not the mechanics of vision, which intrigued many physicists. We pause now to reflect on the behavior of light, and we'll defer the discussion of its nature to later chapters. As betrayed by our fascination, we are still searching for light's elusive definition.

2.2 All Things Equal

Though early philosophers lacked a fundamental understanding of how we see things and how to test their theories of light, several Greek and Arabic philosophers forged ahead in developing rigorous geometric theories about light and its interaction with the everyday world. Euclid is credited with developing the first and most sensible law of optics, called the *Law of Equal Angles*. Simply stated, the Law of Equal Angles says it doesn't matter what light is made of. If light shines on a surface at an angle, it will bounce off the surface at exactly the same angle.

If light comes in at 67°, it will leave at 67°. If it comes in straight down on the surface at 0°, it will bounce right back up. Figure 2.2 displays the simple relationship between an incoming ray of light, the surface it encounters, the *normal* (an imaginary line perpendicular to the surface), and the outgoing ray.

As anyone who has used a hand mirror to blind their little sister can attest, the Law of Equal Angles holds in every situation, even when the surface is not flat

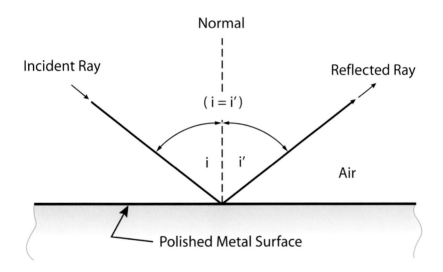

Fig. 2.2 The Law of Equal Angles. An incoming light ray incident on a surface at an angle *i* to the *normal* will reflect from the surface at an angle *i'*, which is equal to *i*. The normal is a line perpendicular to the surface at any point.

like a mirror. Have you ever tried frying an ant on the sidewalk with a curved mirror?[2] It's the Law of Equal Angles that focuses the sunlight to a point (or, if the mirror is turned around, sends all the light rays away from each other). Let's look a little more closely at how light is reflected from various surfaces.

2.3 From the Looking Glass

The Law of Equal Angles can be used in many different ways and is beautifully evident in nature. The most obvious use is the plane mirror. Like most useful tools humans have invented, the mirror's early history is lost in the fog of time; the first mirrors were simple pools of water. Pre-dynastic Egyptian cultures used slabs of slate and gypsum for mirrors up to 6500 years ago, as shown in Fig. 2.3. Ornate bronze and copper mirrors have been found in India dating to 3000 years BCE. Archimedes and the Greeks understood many of the properties of mirrors

Fig. 2.3 A hand mirror from Egypt, c.1400 BCE. (From Roche et al. *Mirrors*, Rizzoli, NY, 1985)

---

[2] The authors neither condone nor admit to such an act, but the result is quite interesting.

and (according to legend) used polished and curved metal mirrors to focus the Sun's light onto enemy ships advancing on Syracuse, burning an entire fleet in the process. As a scientific tool, the mirror's heyday came in the eighteenth and nineteenth centuries: Newton borrowed an idea from seventeenth-century astronomer James Gregory and built the reflecting telescope, an improvement on long, unwieldy *refracting* telescopes. Jean Bernard Foucault in the 1850s and, later, Albert Michelson, used rotating mirrors to measure the speed of light.

In terms of its physical properties, a plane mirror is very simple: any flat reflecting surface will do. Its optical properties are just as simple: light is incident on the reflecting surface and is reflected at the same angle at which it arrives. The best plane mirrors are those that use highly polished silver (or aluminum) as the reflecting (or *first*) surface, since silver reflects almost 100% of the visible light incident on it. Silver's downside as the first surface, however, is its tendency to tarnish and scratch easily. Most commercial mirrors are made with a glass plate over the silver to prevent tarnishing and scratching. The layer of glass, however, degrades the image somewhat because glass also reflects light.

Have you ever looked at yourself as you walked by a storefront window? The photograph of Misty (Fig. 2.4) demonstrates the reflectivity of glass, which is

Fig. 2.4 A glass window reflects 4% of the light incident upon it. If the light from the object hits the window at an oblique (i.e., large) angle, more than 4% of the light will be reflected. (Photo by J. Phillips)

Fig. 2.5 The Hubble Space Telescope mirror, a concave mirror 2.4 meters (8 ft.) in diameter and weighing 1800 lbs. The curved surface of the mirror creates the large image of the technician. (Courtesy of NASA/STScI)

about 4% for most angles. At the *oblique* angle the photo was taken, even more reflection than normal occurs due to the light from the object (Misty, in this case) grazing the surface of the glass. At very large angles of incidence, as much as 100% of the light can be reflected at the surface.[3] You can also see instances of *grazing incidence* in smooth ponds or lakes and in wet streets.

Grazing incidence aside, if you look closely at a standard bathroom or decorative mirror, you'll see a dim, ghost-like secondary image of yourself in the glass. This secondary image is caused by the small amount of light reflection from the first surface of glass. For this reason, astronomers and aerospace engineers who need perfect light reflection without secondary images will use mirrors whose reflecting surface is only polished silver without the glass protection. The Hubble Space Telescope's mirror (Fig. 2.5) is a mirror of this type.

But the Hubble mirror is more than just a fancy reflector. At 2.4 meters in diameter and a cost of $300 million, the Hubble mirror is one of the largest and

---

[3] In Chapter 3 we'll explore the concept of total internal reflection in substances such as water and diamond.

most expensive mirrors ever made (Chaisson 1998). More importantly, because its primary purpose is to investigate celestial objects billions of miles away, the Hubble Space Telescope, like our ant-frying mirror, needs to reflect light in order to focus it. In the next chapter we will look at how lenses use refraction to focus light. For now, however, we're interested in finding out how mirrors use reflection to focus light, and the only way to do this is to make the mirror curved.

2.4 The Curved Mirror

Let's face it, plane mirrors can be plain boring. Curvaceous mirrors have more fun. (Who hasn't had fun looking in the wavy mirrors at a carnival?) Many people use curved mirrors every day if they drive a car with a side view mirror that proclaims, "Objects in mirror are closer than they appear." Have you ever looked into a spoon? Polish one and you've got a wonderful two-sided curved mirror. The cosmetic mirror in your bathroom that lets you look so closely at your skin is a simple concave mirror, focusing light to create a magnified image of your face. The cosmetic mirror has a reflecting surface that curves inward, away from the object. As you might expect, a *convex* mirror bends outward, toward the object.

Other than physical traits, the major difference between concave and convex is the directions in which they reflect light. Since concave mirrors are curved inward, they *converge* light so that it's focused toward a point on the inside side of the curve, as in Fig. 2.6. Convex mirrors, as we'll see, *diverge* light — the opposite of focusing. But if mirrors obey the Law of Equal Angles, how can a curved mirror reflect light rays so they all end up at (or near) a focal point? Doesn't the focal property disobey the Law of Equal Angles?

If we look close enough at a curved mirror's surface, we can imagine the mirror as hundreds of flat surfaces lined up end to end, each placed at a slight angle to its neighbor. At each point on the surface, the mirror still obeys the Law of Equal Angles because, to a light beam, each point is just like a tiny flat mirror. Figure 2.6 shows us a close-up view of what happens to light beams that encounter these tiny flat mirrors: each light beam hits the surface and reflects, obeying the Law of Equal Angles. The trick is that each surface is angled slightly differently than its neighbors, so each light beam heads off in a slightly different direction. Viewed as a whole, all the light beams reflect differently from their neighbor, but all converge to a point, or diverge as if they were coming from a point.

The Hubble Space Telescope mirror in Fig. 2.5 is a *concave* mirror, one that focuses light to a point somewhere in front of the mirror's surface. Where exactly that point will be depends on the shape of the surface. Figure 2.7 shows the (exaggerated) difference between two concave mirrors that reflect light to the

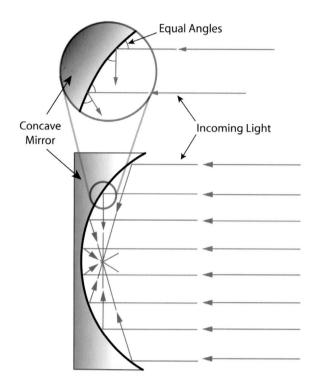

Fig. 2.6 Each point of the concave mirror is like a tiny flat mirror tilted slightly from its neighboring points. Each ray of light is reflected at an angle equal to its incoming angle.

focal point. In Fig. 2.7(a), the mirror is not sharply curved, so the light beams come to a focus far away from the mirror. When the mirror *is* sharply curved, as in Fig. 2.7(b), the light beams converge to a focal point much closer to the mirror. It is this difference that causes the differences in magnification between curved mirrors. As we'll see in Chapter 3, it is the difference between lenses as well.

2.5 Shaving and a Spoon

When we flip a spoon over, we find we're upside down and our heads have shrunk. Welcome to the weird world captured in a convex mirror. Figure 2.8(a) exhibits how light incident on the surface of a convex mirror will behave: instead of reflecting inward, the light rays reflect away from each other — they *di-*

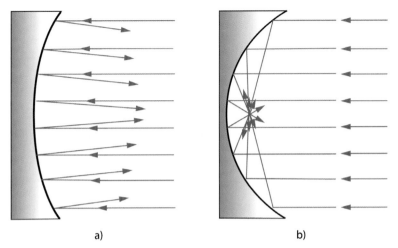

Fig. 2.7 Two concave mirrors reflecting incoming parallel light rays to a focus. **(a)** A shallow curved mirror has a focal point far from the mirror surface. **(b)** A steeply curved mirror has a focal point close to the mirror.

verge. Similarly to concave mirrors, the more a convex mirror is curved, the "faster" the light will diverge.

The tendency of light to diverge from a convex mirror makes for quite bizarre images. In Chapter 1 we found that the paths of light are reversible; a ray of light traveling in a certain direction will follow the exact same path backward if we reverse the process. Taking this concept to mirrors (and later, to lenses), we can imagine reversing the light rays in Fig. 2.8(a), having them come toward the mirror from divergent places. The light reflects off the mirror and leaves in an orderly fashion, in parallel as in Fig. 2.8(b). In this way, we see that objects far outside the view of a flat mirror can be seen quite clearly in a convex or wide-angle mirror. We mentioned the passenger-side mirror of an automobile: these are convex mirrors, which allow the driver to see objects outside the range of what a standard plane mirror would show. The downside is that it makes those objects look far away in the mirror, thus the warning written on the mirror.

One of the notable differences between images seen in a concave mirror versus a convex mirror is that the image sizes are generally not the same. Concave mirrors produce large images that are often upside down, depending on both the location of the object and of the observer relative to the focal point of the mirror. You can see this effect in a makeup or shaving mirror at home. Try viewing your image in the mirror when you stand very close to it (as you would when shaving or applying makeup). Then try looking at the mirror when standing a few feet away. Up close, you are standing between the focal point and the mirror, and your image is larger, or magnified. Far away, you are outside the focal point, and the image is still larger, but you are upside down, or inverted. As an

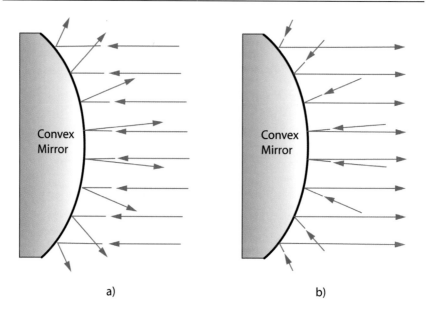

Fig. 2.8 (a) Incoming parallel light rays reflect from a convex mirror and bounce off in many directions — they diverge. **(b)** In reverse, the light rays come from divergent places and leave in parallel.

added thrill, see if you can find the point where the object (you) passes through the focal point of the mirror. What happens to your image at this point?

If you have a polished spoon, try looking into the backside of the spoon (the convex side). If you compare the images from the bathroom mirror and the spoon, you will see that the spoon image is smaller and right-side up, but that you can see more of the room. This will always be the case for convex mirror images, because these images are made by light that seems to come from a point on the other side of the mirror. The rays from the convex mirror never cross, so the image is always upright and smaller than the object. The advantage, though, is that more of the light from the space surrounding the object comes into view. With the flip of a spoon we can see a grain of sand or the world.

2.6 The Rough Edges

Most things are not polished smooth like a mirror. We know this without even touching them. How? For starters, we can see them. If every object was polished smoothly, we'd live in a frightening and confusing world. Everything would be like clean stainless steel. The Law of Equal Angles would rule absolutely. Need

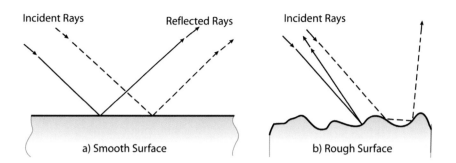

Fig. 2.9 (a) With specular reflection, incident light rays that are parallel to each other will be reflected in parallel. Often we can see images of distant objects in such a surface. **(b)** With diffuse reflection, incoming parallel beams are reflected in different directions. It's difficult to see objects reflected in a matte surface.

to grab that coffee mug? Don't look at the mug itself — look at the reflection in the mug of your hand approaching it. When your hand and the image of your hand come together, you've reached the mug. Often you wouldn't even be able to tell the difference between the thing you wanted to look at and its image. Thankfully, the world is not polished smooth, and we can see actual objects, not just their images. The property that lets us see the world is more a property of objects than it is of light, but it is important to our discussion nonetheless. When light encounters a smooth surface, such as in Fig. 2.9(a), every reflection is perfect at each point, and light rays that were parallel before they struck the surface will be parallel after they reflect. This is known as *specular* reflection. Every light ray from an object gets reflected perfectly, and we can see the image of that object in the smooth surface.

Conversely, *diffuse* reflection occurs when light encounters a surface that is uneven. This is the case in Fig. 2.9(b). Light rays encounter the rough surface and are reflected in many directions. Notice that at each point the reflection still obeys the Law of Equal Angles. However, each reflecting surface is not parallel to the next. Since every light ray from the object is not reflected in parallel, the object's light that reaches our eyes is not coherent, so we cannot see an image of it. The light is diffused and we only see the reflecting object itself, not images of other objects in its surface.

Photo labs will ask if you want your photographic prints glossy or matte. You'll notice that the glossy prints are smooth and reflect surrounding light quite well, while matte prints are rough and keep light reflection to a minimum.

One of the earliest forms of photography, the daguerreotype, relies heavily on the reflection of light. Let's shift our focus from the general properties of light and take a closer look at the beautiful art and science of daguerreotypes.

3 Daguerreotypes: Light Captured

3.1 A Race to Capture Light

In the early 1830s, a flamboyant, middle-aged artist stumbled upon a process of capturing light that changed the worlds of art, journalism, and memory. Louis Jacques Mande Daguerre, a French painter and designer, discovered a way to fix certain chemicals to a metal plate after they had been exposed to light. These first "photographs" were in a sense both matte and glossy at the same time. He was not alone in his discovery: William Henry Fox Talbot and Joseph Nicephore Niepce had both independently come upon the same result using slightly different chemicals. But it was Daguerre who most immediately applied his process to capturing images and who opened the door for scores of others to create objects that, though now obsolete, are considered by some to be the most beautiful and haunting works to come from the field of photography.

To give due credit, it was Niepce who made the first photograph, a scene of his garden outside *Chalon-sur-Saone* in 1826 or 1827. His process involved exposing a silver pewter plate covered with *asphaltum*, a chemical that hardened on exposure to light and was easily washed away when unexposed and still soft.

But his photograph was not reproducible and was quite impractical (only his original photograph still exists). Furthermore, he failed to publish his findings for several years, allowing others (such as Daguerre and Talbot) to independently discover alternative processes for capturing and developing *point vue* images of nature.

It was his inability to draw that led Talbot in 1835 to independently invent the modern photographic process. He wanted to capture an Italian landscape in lithograph — the process of painting ink on stone and transferring the image to paper — but in frustration he threw up his hands and vowed to find an easier way to "cause these natural images to imprint themselves durably and remain fixed upon the paper." His original pictures were rendered as negatives and required at least an hour of exposure time. In 1840, several years after Daguerre, Talbot discovered that a *latent image*, requiring only a few minutes' exposure, could be developed into a *calotype,* or positive photograph. Eventually, Talbot's paper photographic process would emerge the more practical (and hence,

Fig. 3.1 Vintage antique 1895 daguerreotype of boy.

profitable) endeavor, but in the mid-nineteenth century, Daguerre had beat Talbot in the publishing race and was commercializing on his invention, the daguerreotype, which had captured the world in its three-minute exposure.

Daguerreotypes are photographs' long-lost older brother. They both come from a similar general process, but their chemical and physical properties are slightly different. At a distance they look similar, but up close they're night and day, mostly regarding how the images are created and viewed. In this chapter we will look at the general principles behind the formation of photographs and daguerreotypes, and we'll see why they are so fascinating in the context of reflection.

3.2 Tripping the Light Fantastic

At first glance, daguerreotypes look like traditional black-and-white photos. Surely many people have sadly discarded or ignored those old photographs sitting in the trunk in the attic that weren't really photographs at all! Upon closer inspection, daguerreotype images can be very clear and have extremely high *resolution*, especially coming from such an old process. Both the daguerreotype and modern photographic processes use similar chemicals to reach a similar outcome — the formation of a permanent image on a portable medium. The differences become evident when we go beyond the cursory glance. For one thing, daguerreotypes are made of metal; photographs are made of paper and chemicals.[1]

Traditional photographs and daguerreotypes alike use a base of silver crystals that are altered when exposed to light. Modern photographs use an emulsion of silver chloride (AgCl) with plastic as the film base; it's somewhat like a layer of silver mayonnaise on a strip of plastic. When the film is exposed to light, places where the light strikes a silver crystal are partly converted into metallic silver. If an area on the film is exposed to a lot of light, many crystals will contain a nugget of metallic silver. Where there is little light exposure, the silver will remain in compound form.

A chemical process is used to reduce (or *develop*) the silver chloride into silver. As the developer works, the silver crystals containing the metallic nuggets are converted into pure silver. A separate process (using a chemical *stopper*) then stops the chemical development so that those crystals not exposed to much light will remain silver chloride crystals while the exposed areas will be metallic silver. A final chemical process (called *fixing*) washes away the remaining unexposed silver crystals, leaving only the metallic silver behind. This type of chemical development process produces a *negative* image because the exposed areas (metallic silver) are dark, and the unexposed areas are transparent. If you look through an exposed and developed film negative, you can see dark spots where the film has been exposed, and light spots where the film is unexposed and the light passes right through. An illustration of the major steps in the photographic process is shown in Fig. 3.2.

Daguerreotypes are in some ways both simpler and more complex than photographs. They are simpler because there are fewer steps in the development process. They are more complex because the optical properties of daguerreotypes (i.e., reflection and absorption) are subtle and non-intuitive. In his process, Daguerre laid a thin layer of silver over a copper plate and polished it. Then he put the plate in a vapor of iodine, which bonded with the silver to make crystals of silver iodide.[2] Exposure to light yielded an intermediate step similar to photographs: exposed areas contained crystals with metallic silver, while unexposed areas just had the crystals. What Daguerre did next put him in the art history and science books: he found out quite by accident that if he immersed the exposed plate in a vapor of mercury, the mercury would form large crystals around the metallic silver nuggets. This was the precursor of the modern *development* stage.

---

[1] Fox Talbot also invented the positive-negative process, whereby light is shined through a negative onto photographic paper to yield a positive image.

[2] You can already imagine that reproducing a daguerreotype is difficult.

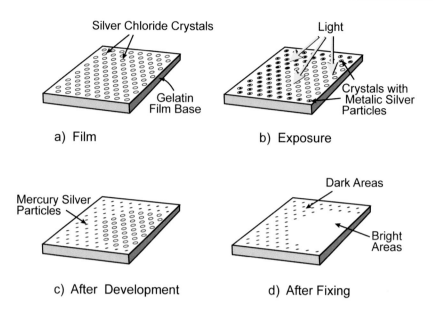

Fig. 3.2 The photographic development process. **(a)** A plastic film base is coated with an emulsion of silver chloride in a gelatin paste. **(b)** During exposure, the silver chloride crystals exposed to light form a tiny nugget of metallic silver. **(c)** In the chemical development process, the metallic silver "seeds" are grown into larger silver particles, or grains. **(d)** Washing the film with fixer removes the unexposed silver chloride and leaves the metallic silver grains.

Like water condensing on a cold can of soda, the droplets of mercury grew larger and larger around the silver nuggets. Eventually an image formed, but in Daguerre's case, the areas exposed to light came out white and the unexposed areas came out dark — a positive image.

Daguerre's next step was also visionary: in order to keep the plate from being exposed again and again to light (remember, the development process takes place in a darkened room), he had to *fix* the mercury-silver crystals to the plate, but remove the silver iodide crystals, those that are sensitive to light. This he did with sodium thiosulfate, the chemical used as a photographic fixer to this day. He was left with a plate of polished silver metal that contained tightly packed mercury-silver crystals in the bright, exposed areas of the plate. The dark areas were simply the polished silver that remained after the other chemicals had been washed away by the fixer. The steps in Daguerre's process are shown in Fig. 3.3, but the properties of reflection tell us why this metal photograph looks so similar and yet so different than later paper photographs.

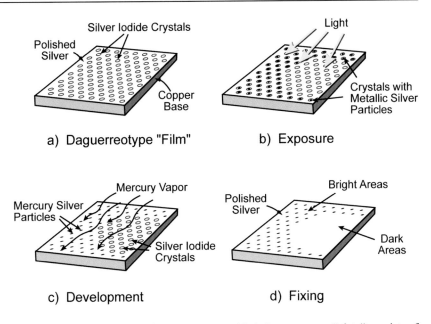

a) Daguerreotype "Film"   b) Exposure

c) Development   d) Fixing

Fig. 3.3 The daguerreotype development process. **(a)** A daguerreotype "plate" consists of polished silver overlaying a copper base. The plate is bathed in iodine to form silver iodide crystals. **(b)** During exposure, the silver chloride crystals exposed to light form a tiny nugget of metallic silver. **(c)** The plate is bathed in mercury vapor to "develop" the exposed nuggets into mercury-silver particles. **(d)** Washing the film with fixer removes the unexposed silver chloride and leaves the mercury-silver grains.

3.3 It's all in the Reflection

If you're lucky enough to own a daguerreotype or you have a relative who has kept one, place the daguerreotype on a table next to a black-and-white photograph. You might immediately notice that the *resolution*[3] of the daguerreotype image is equal to, if not better than, that of the photograph — an amazing feat for nineteenth-century science. You may notice that the image is easier to see at some angles than at others; in fact, the daguerreotype image may disappear altogether at some angles. Now notice that the photograph looks basically the same at any angle. Which image do you think employs diffuse reflection and which

---

[3] How finely you can see details.

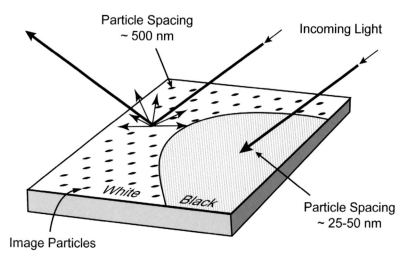

Fig. 3.4 The reflection properties of modern black-and-white photographs. In the dark areas of the photograph, silver particle spacing is small and light is quickly absorbed. In the bright areas, particles are spaced far apart and light either passes through (in the case of transparent film) or is diffusely reflected (in the case of photographic print paper).

uses specular reflection? Figure 3.4 illustrates how the photograph reflects light: in the dark areas of the photograph, the silver nuggets are spaced very closely (25–50 nm), so light is reflected from particle to particle and is absorbed very quickly. In the white areas of the photograph, the silver particles are spaced much farther apart (~500 nm), and so the light can pass right between the silver nuggets. This means that, on a photographic film negative, the light will pass through the negative in these unexposed areas. (See for yourself!) On the photographic print, the underlying paper reflects the incident light diffusely, in all directions, and so the region looks white.[4] The silver particle spacing is everything in modern photographs.

In the daguerreotype, we're not so lucky. Not only do we have to deal with particle spacing, we also have to think about viewing geometry. Think of a big, fluffy storm cloud in a darkened sky. If the cloud is in the foreground of other darker clouds, it will look almost white. But if the cloud is in the foreground of other *lighter* clouds, it will look dark gray, almost black. Darkness is relative. Our visual system compares the brightness and color of objects when interpreting visual input. In this way, black can take on different values, especially when considered in its surroundings of other darker or lighter objects. The daguerreo-

---

[4] We'll see why diffuse light scattering causes objects to appear white in Chapter 7.

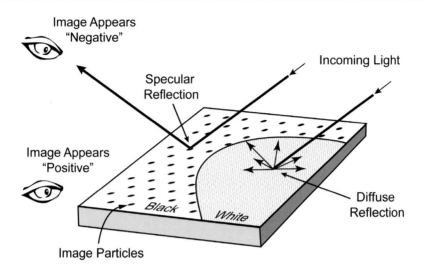

Fig. 3.5 The reflection properties of daguerreotypes. In the bright areas of the daguerreotype, particles and particle spacing are small and light is diffusely reflected. In the dark areas, large particles of mercury-silver are spaced far apart, so light is specularly reflected from the polished silver underneath. If the daguerreotype is viewed at an angle such that the light source is reflected in the plate, the image is reversed.

type exhibits this ambiguity. From most angles, the daguerreotype appears positive: exposed areas appear white, and unexposed areas appear black. But at certain angles, the exact opposite occurs: exposed areas appear black, and unexposed areas are white. Why is this happening?

Figure 3.5 shows why daguerreotypes are wonderful examples of specular reflection. We know that in the exposed areas of the image, mercury-silver particles formed after exposure and development. The mercury-silver image particles are extraordinarily small and very tightly packed; so much so that they interact very well with visible light.[5] White light is scattered from the exposed areas of the daguerreotype very efficiently and in all directions.

The unexposed areas of the daguerreotype were washed of all silver-iodide particles by the fixer, and essentially only the smoothly polished silver base remained. When ambient light is incident on the polished silver, the light is *specularly* reflected (polished silver reflects more than 95% of visible light). This is where viewing geometry enters the picture.

Daguerreotypes are exhibited in black, velvet-lined cases for two reasons. The first is to protect the priceless objects. Even though they're made of metal,

---

[5] Whose wavelengths are very near the spacing of the mercury-silver particles. We'll learn about wavelengths in Chapter 6.

a)   b)

Fig. 3.6 A daguerreotype illuminated **(a)** under ambient light reflects a positive image **(b)** under ambient light and viewed at an angle so that the light source is reflected in the image. A negative image is observed.

daguerreotypes are much more fragile than standard photographs; the mercury-silver crystals are deposited directly onto the plate and can be rubbed off with a light brush of a finger. (Never remove a daguerreotype from its glass case.) The second reason for the black velvet is optics. Under ambient white light without the black case, a daguerreotype image is difficult to see because of the unexposed, polished silver areas that reflect the light so well. But put the daguerreotype under light-absorbing black velvet, and the unexposed silver reflects no light — a wonderful contrast appears between the exposed, light-diffusing image areas and the unexposed, light-reflecting (but now black) areas. The black, unlined daguerreotype of Fig. 3.6(a) shows the contrast between the image and the background.

If you hold a daguerreotype without the black velvet case and position it at such an angle that the light source is reflected in the daguerreotype, as in Fig. 3.6(b), a transformation occurs: the image is reversed — a negative. Like the sunlit cloud against the dark sky, we see that the light reflected in the silver is much brighter than the diffuse light from the exposed areas that now look black. The properties of reflection and a simple change of viewing angle allow us to switch, like a primitive hologram, between the positive and the negative, the seen and the unseen.

Despite its inherent beauty and its popularity in the 1840s and '50s, the daguerreotype process eventually succumbed to more robust and practical film- and paper-based processes pioneered by scores of new photographic aficionados. None of these new techniques relied quite so heavily on the properties of reflection in perceiving the image, but they did rely on lenses and their ability to produce clear, sharp images. We will explore photography more in Chapter 5 when we look at lenses and their role in producing images. But for now, we leave the daguerreotype detour and discover the properties of light that make lenses work.

4 The Refraction of Light

4.1 From Galaxy to Fish

Whether it's illuminating a photographic plate or passing through a window, light changes materials and, in return, it is changed itself. It may travel for billions of miles without encountering one obstacle, and in its nearly infinite journey strike the tiny lens of a telescope sitting in an amateur astronomer's backyard. In that encounter, the light is changed. Its speed is decreased and its path is altered. If an eye or a camera is at the end of the telescope, the light gives up the ghost and alters the chemicals in the eye or the film. Energy is transferred, information is passed on, a galaxy is recorded.

Backward across the span of time, the light from the galaxy midway in its journey to lonely Earth, a shaft of sunlight glints off the scales of a fish, crosses from the stream into the warm, Paleozoic air and into the eye of a not-so-successful fisherman. With that flash of silver, the fisherman realizes he's failed to catch a fish because the fish aren't where he thought they were. He wonders if perhaps the water goddess cast a spell on the water to fool him. Or perhaps something happened in the flash from the water to the air. Or maybe the fish is actually closer to him than it seems. He suddenly realizes that the fish appears to be in a different spot than it really is. He quickly reaches for empty river bottom and comes up with dinner.

4.2 Altering the Speed of Light

Not so many years after that pioneering fisherman, the Greek mathematician Euclid described an observation usually attributed to Aristotle: the lower portion of a stick in an empty water pail, viewed at an angle, becomes visible when the pail is filled with water. Today, we can look at a pencil in a cup of water or see our legs dangling in a pool and understand that they're not really broken. But why do they seem bent? And what does this have to do with fishing?

Fig. 4.1 The light from this galaxy travels at the exact same speed in its 300 billion-kilometer journey across deep space. Only when it enters our atmosphere and encounters a camera or a telescope lens does it slow down. (Courtesy NASA/STScI/AURA)

Refraction, the bending of light as it crosses a boundary, was the next step in understanding the properties of light. After he gazed on his reflection in the smooth surface of the lake, the fisherman tried to understand what he saw below the surface. Soon after, he did. But it took a few millennia for his descendents to explain it.

Two things happen simultaneously when light encounters a transparent medium such as glass, water, or a layer of heated air. First, part of the light is reflected from the surface and second, the transmitted beam is refracted. As we

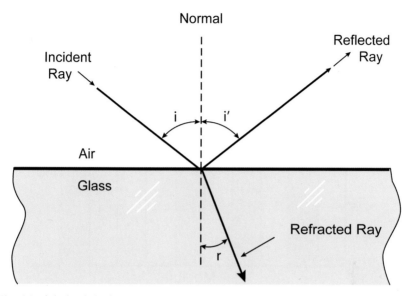

Fig. 4.2 Light in air incident on a glass surface where it is partly reflected at the surface and partly transmitted into the glass. The direction of the transmitted ray is changed at the air/gas interface. The angle of refraction, r, is less than the angle of incidence, i.

saw in the previous chapter, 4% of incident light is reflected from a glass window. As the incoming angle increases toward the horizontal, that percentage goes up quickly. The rest of the light is transmitted through the surface and into the material. Of the two fates of the incident beam, the transmitted component of the light is what interests us here: the transmitted light changes direction at the surface — it *refracts*. As shown in Fig. 4.2, the transmitted, or refracted, beam changes direction at the surface and deviates from a straight line path of the incident light ray.

Why? The change of direction of light as it passes from one medium to another is associated with a change in the *velocity* and *wavelength* of the light. (We'll discuss waves more in the next section.) In uniform, transparent liquids or solids such as glass, water, Lucite, oil, etc., these changes occur at the surface and not throughout the material. In gases and layered liquids and solids, however, the light refraction can occur gradually as the light penetrates layers of greater or less density.

Visible light in air slows down when it enters a medium such as glass, like a skydiver experiencing an upward gust of wind. The velocity of light in the glass decreases to three-quarters of its velocity in air. In other materials the velocity decrease can be even more substantial. For example, in linseed oil, the velocity decreases to two-thirds of its velocity in air. Figure 4.3 displays a bar chart of

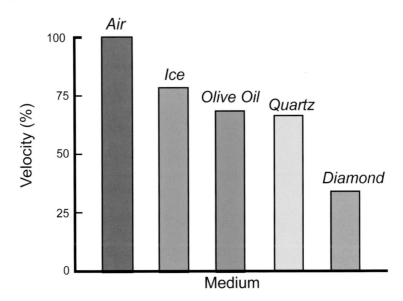

Fig. 4.3 Bar chart of the velocity of light in various media. The value of 100% refers to the velocity of light in vacuum.

the velocity of light in different media. The 100% value is the velocity of the light in vacuum (e.g., outer space or a laboratory environment). In these environs, light truly experiences the speed of light. For air, the velocity is 99.97% of the speed in vacuum. In diamond, the light slows to nearly half its value in air. We'll reveal more about diamonds later in this chapter.

Light slows down, but why should that make it bend? The answers are in the phenomena of waves and their interaction with matter. As we'll see, sometimes it is easiest to view light as vast undulations carrying energy from one place to the next. At other times, light acts like a simple particle — a photon bouncing here and there, a tiny packet of energy blinking in and out of existence. The theory of quantum mechanics tells us that both viewpoints are valid and correct. It is up to us, the experimenters, to decide when to choose one view over another. With refraction, we'll adopt the wave viewpoint.

4.3 The Light Brigade

Isaac Newton did not like waves. Waves did not explain why light traveled in a straight line. For him, light consisted of *corpuscles*, tiny particles of light that were easily subjected to the forces of nature. Newton postulated that every optical phenomenon was explainable using corpuscles, and they fit in perfectly with his worldview: forces, Newton's *ne plus ultra*, could act on particles, not on waves. Newton's tremendous stature in the scientific community of the eighteenth century allowed his corpuscular theory of light to continue for nearly one hundred years. Two of his contemporaries, Hooke (his bitter rival) and Huygens, laid the foundation for the wave theory of light, which dominated most of the next century.

In the mid-1600s, when he wasn't rebuilding London after the Great Fire or founding cell biology, Robert Hooke observed colored fringes of light while looking at thin films of mica and concluded light traveled in waves. Several years later, Christiaan Huygens proposed that a light source emitted spherical waves of light, each point of which generated new spherical waves, thus propagating the wave forward. He derived all the previous optical laws from his wave theory: straight line motion, reflection, and refraction. He also solved several optical problems that couldn't be explained by the Newtonian corpuscular theory.

One problem that wouldn't go away for the wave theorists was Newton's complaint that he could *hear* a friend around a corner, but not *see* him. If light is a wave, he pondered publicly, one should be able to see a sound as a wave around a corner. In 1803, Thomas Young looked around the corner and sounded

Fig. 4.4 A portrait of Isaac Newton. (Courtesy Library of Congress)

the death knell of the corpuscular theory. Young's experiment, in which alternating fringes of dark and light bands emerged from two slits illuminated by a source of light, was so easily explained by wave theory that it set the tone for not only nineteenth-century optics, but for all of physics as well.[1] A.J. Fresnel's elegant mathematics and Michael Faraday's magical experiments drove the wave theory from scorn into the realm of the ether, while Maxwell and Michelson unified light, waves, electricity and magnetism and abolished the ether altogether. At the turn of the century, Planck and Einstein quantized the corpuscle (now, photon) and tied it to matter with the speed of light as its string with Einstein's famous $E = mc^2$.

4.4 The Properties of Waves

Waves surround and pierce us. Most of them are unseen: disturbances in air as sound, traveling electric and magnetic fields as light. On the surface of liquids, we see them spread outward from a dropped stone or crash violently against a shore. All waves (like everything else, really) are just forms of energy. As such, though disparate in their manifestations, all waves share similar properties. Every wave has peaks and troughs: the top-to-bottom distance of the wave trough and peak is its *amplitude*. Every wave has a speed, a wavelength, and a *frequency*. Speed tells us how quickly the wave moves. Wavelength is a measure of the distance from peak to peak or trough to trough of successive waves. Frequency tells us how often the peaks (or troughs) pass if we are standing in one place watching the wave go by.

An astute observer might notice that speed, wavelength, and frequency are all closely related: wavelength is a distance, frequency is the inverse of time (waves *per second*), and the wave speed is the product of the two, just like the velocity of a vehicle — a distance per unit time. On the Earth, waves move at many different speeds. Water waves move slowly, sound waves move faster, and light waves the fastest.

As mentioned in Chapter 1, all electromagnetic (light) waves move through space at the same speed, 650 million miles per hour, regardless of whether they're long-wavelength radio waves or short-wavelength x-rays. However, the energy of a wave is intimately (and inversely) linked to its wavelength, so x-rays with small wavelengths have high energies, while radio waves are not very energetic. (This is why x-rays can be damaging to living tissue, while radio waves are not.) Since the wave energy is associated with the wavelength, it must also be linked to the wave speed. Usually, as light is just going along minding its

---

[1] We briefly discuss diffraction and its effects below and in Chapter 7.

Fig. 4.5 Water and waves on a pond. The distance between the peaks (or troughs) of the waves is the wavelength. The number of waves to pass a point in some amount of time (the standard is one second) is the wave frequency. (Photo by Steve Beeson)

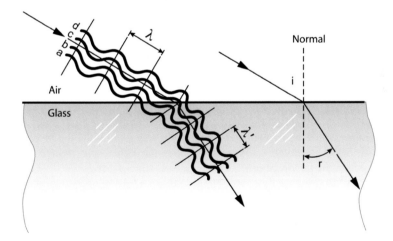

Fig. 4.6 Light waves of wavelength λ incident on glass change direction and wavelength when transmitted into the glass.

own business, its speed does not change. In the cold vacuum of space or hurtling through the atmosphere, the light wave energy is determined only by the wavelength.

But in the transition from one medium to another, light waves must maintain the same energy as they change speed across the boundary. To compensate and keep the energy constant, its wavelength must change. If the wave slows down, the wavelength must decrease. If it speeds up, the wavelength must increase.

We illustrate this concept in Fig. 4.6 by representing incident light as parallel waves with a uniform wavelength, λ, between the wave fronts, the lines connecting the peaks. As the light enters the glass, the wavelength changes to a smaller value, λ'. Wave a passes the air/glass interface and slows down before b, c, or d arrive. The break in the wave front across the interface occurs when waves a and b have entered the glass, slowed down, and changed direction. At the next wave front in the glass, all four waves are traveling with the same velocity and wavelength λ'. More importantly, the entire wave has changed direction.

4.5 On the Beach

Perhaps a better way to visualize a wave encountering a medium of different density is to go to the beach. Imagine a long line of people running into the ocean, one after another. As the first few people run into the water, they're slowed down because it's harder to run in water. Therefore, they bunch up and stay bunched up as they run through the water. When everyone in line has entered the water, there's a line of people all running in the same direction, but the line is shorter and the runners are bunched closer together, as in Fig. 4.7(a). As they run back to the beach, the first few people clear the water and run faster. Eventually, everyone clears the water and runs at the original pace with the original spacing between persons.

In this analogy, we represent the light wave as the whole line of people and the crests of the wave as the individual persons. The distance from one person to her neighbor would be the wavelength, and the water would be the medium into which the wave is traveling. Why, then, does the light wave change direction when it enters the new medium?

We can extend our analogy and imagine two lines of people running into the ocean from the beach. In Fig. 4.7(b), the lines are close together, and each person in a line is holding hands with a person in the other line. Each pair can be considered a wave front, like the lines joining the crests in Fig. 4.6. Waves a, b, c and d in that figure represent each long line of runners. As soon as a runner enters the wave, he slows down. In order to maintain the one-to-one relationship with the other line (and to keep holding hands), both lines must turn when they

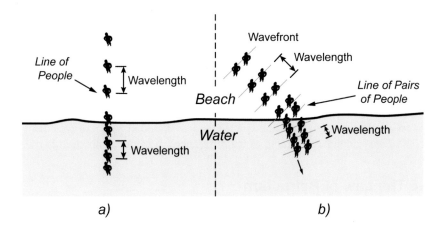

Fig. 4.7 In the beach analogy, a line of people (representing a wave) runs from a sandy beach into the ocean. **(a)** The distance between runners (the wavelength) decreases when they hit the water **(b)** With a pair of runners representing a wave front, the "wave" slows down and turns when it hits the water.

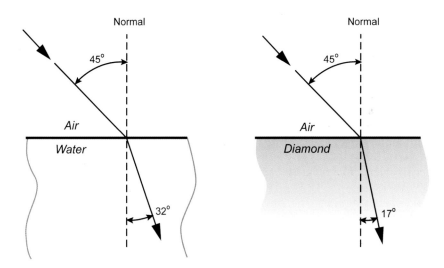

Fig. 4.8 Light incident at 45° on water and diamond. The angles of refraction (32° for water and 17° for diamond) depend on the optical properties of the materials. The reflected components are not shown.

enter the water, and the wave front bends. When the runners turn, they turn toward the normal — the imaginary line that runs perpendicular to the interface between the two media (the water and the beach); a pier is an example of a normal.

So the two lines (the waves) must bend toward the normal when they hit the water. The greater the change in velocity and wavelength, the greater the change in direction. Figure 4.8 shows the change in direction for light in air incident at 45° on water with a refracted angle of 32° and on diamond with a refracted angle of 17°. These angles correspond to the differences in velocity shown in Fig. 4.3.

4.6 The Law of Refraction

We are now ready to discuss the Law of Refraction, discovered by Willebrod Snell van Royen and quantified by René Descartes: When light passes into a medium in which its velocity decreases, the light will bend toward the normal.

Fig. 4.9 Portrait of René Descartes (Courtesy Library of Congress)

When light passes into a medium in which the velocity of light is greater, the light will bend away from the normal. The amount that the light bends depends on the incoming angle and the change in velocity that the light experiences. Mathematically, the ratio of the sines of the angles of the incoming and refracted rays is equal to the ratio of the velocities of light in the two media. Though it is Snell's name attached to the Law of Refraction (except in France), it was Descartes who first offered a rigorous mathematical derivation of the law. With the laws of reflection, refraction, and a few other geometrical odds and ends in hand, Fresnel in the eighteenth century developed a near-complete theory of optical instrumentation, including the properties of diffraction and interference.

But neither Descartes nor Snell formulated a correct theory for the cause of refraction. Each believed (as did just about everyone else, except Pierre de Fermat) that light sped up (rather than slowing down) in a medium. It required the work of the wave theorists, Huygens and Fresnel, to explain why the fish wasn't where it appeared to be. Fresnel will emerge again in the next chapter during our examination of lenses.

4.7 The Refractive Index

Regardless of the medium in which it travels, light travels fast. Because of the large velocities (and hence, large numbers) involved, it's impractical to discuss the optical properties of any material by referring to the light velocity inside of it. Instead, we use an index factor to compare the optical properties transparent materials. The ratio of the velocity of light in a vacuum to the velocity of light in a medium is called the medium's *refractive index*. If we go back to our beach/

Table 4.1 Values of Refractive Index

Medium	Refractive Index
Air	1.0003
Ice	1.31
Water	1.33
Olive Oil	1.47
Crown Glass	1.52
Diamond	2.42

ocean analogy, we can think of the refractive index of the water as something like its density. Air is not very dense at all (its refractive index is 1.0003), so the people run through it quite easily; but when they run into water, which is denser than air (and has a refractive index of 1.333), they slow down and their line bends at the ocean/beach interface. Recall that the speed of light in vacuum is about 650 million miles per hour. If the refractive index for water is 1.333, which represents the ratio of the speed in vacuum to the speed in water, then the light must be traveling about two-thirds as fast in the water, or 490 million miles per hour (225 million meters per second).

Refractive indices are most easily determined from the measured values of the incident angle and the angle of refraction and their geometric relationship. Values of the refractive indices for the media listed in Fig. 4.3 are given in Table 4.1.

The path of light in air incident on and transmitted through a glass plate is shown in Fig 4.10. The angle of the incident ray to the normal is 45° and equals that of the reflected ray. The transmitted ray is refracted at an angle of 28° to the

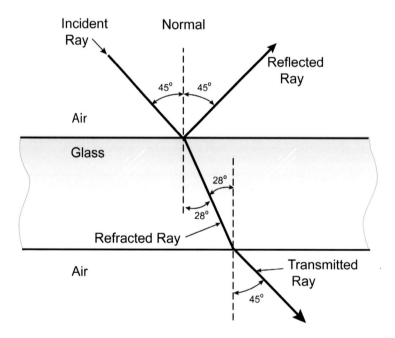

Fig. 4.10 Light incident on a glass plate. The reflected part of the ray is shown along with the incident light path.

normal and exits the glass at an angle of 45° to the normal, an angle equal to that of the incident ray. This explains why, for example, the image we see through a flat-glass window pane is unchanged from the image seen through an open window.

Light that is incident along the normal to the glass plate does not change direction as the transmitted light continues normal to the surface (the air/glass interface). The light is not refracted (i.e., there is no change in angle), but the wavelength and velocity do change, and about 4% of the incident light is still reflected at the air/glass interface.

4.8 Total Internal Reflection

Euclid and Ptolemy both were confused. When they shined light from underwater toward the air, neither could explain why at certain angles the light didn't even come out of the water. You can see this for yourself if you go for a swim in the ocean or a pool and look up at the surface from underwater. If you reverse the light paths in Fig. 4.11 (as we know is valid from Chapter 1), you can imagine being able to see the sky directly above you, but on the edges of your view, the darkness of the ocean depths will be reflected from below.

What is happening? It is the same phenomenon that makes diamonds sparkle so brilliantly. If you've ever seen a natural diamond, you might have been surprised that it didn't sparkle nearly as much as a cut diamond. A jeweler cuts a diamond to take advantage of its light-reflecting properties and to literally trap the light inside the stone.

If light is inside a material such as water, glass, or diamond, that has a larger refractive index than that of the material outside, such as air, and if the light is incident on the interface at an angle greater than a critical value, then the light is entirely reflected back into the material and does not escape. This specific angle (relative to the normal) is called the *critical angle*, and the property is known as *total internal reflection*. For light exiting glass, the critical angle of incidence is 41°. Light in glass that strikes the glass/air interface at 41° or greater all will be reflected back into the glass. Not 4% of the light reflected, but all of it. One hundred percent. The dark shaded area in Fig. 4.11 is the angular region where light is reflected back into the glass. The greater the difference in the two refractive indexes, the smaller the critical angle and the smaller the amount of light that can escape.

For diamond, the index of refraction is 2.42, and the critical angle is 24.5°. That means that every ray of light that enters the diamond must strike the next (inner) surface at an angle less than 24.5° if it's going to escape. If the jeweler has cut the diamond properly, that light will have difficulty escaping and will bounce around inside the diamond for a while, until it strikes the diamond/air

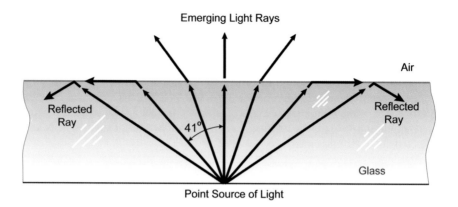

Fig. 4.11 The internal reflectance at an air/glass interface for light rays from a point source in glass. Light rays incident at angles greater than the critical angle (here, 41° for glass to air) do not leave the material and are totally reflected at the interface.

interface at less than the critical angle and escapes. For a brilliant-cut diamond, that escape usually occurs at the facets on the top of the diamond, around the center facet, or table. If you look through the back side of a diamond at a light source, the diamond will appear black because all the light that would normally pass through the diamond has been internally reflected to the front (top) of the stone.

4.9 Diffraction: Newton's Mistake

Light can be bent in ways other than refraction. Like water waves bending around a dock's pylons, light bends around corners and through small openings. This result is impossible to explain by Newton's corpuscular theory but was discovered by Thomas Young in his groundbreaking two-slit experiment, which we will discuss in Chapter 7. *Diffraction* is a property of waves themselves and is independent of the medium in which the wave travels, unlike refraction. The amount of bending is determined by the size of the opening and the wavelength of the wave. Since light's wavelengths are so tiny, the opening needs to be sufficiently small (a few thousand times larger than λ), or the corner sufficiently sharp, in order to effectively see diffraction effects. You can see the diffraction of light for yourself using just your fingers and a bright light source (not the sun). Bring your thumb and index finger next to one eye (keep the other eye closed) and look at a light bulb or the sky with your fingers very close to one another, as if you're holding an ant. The tiny fringes you see in the space between the fingers are the effects of light being diffracted as it passes through the slit you've created. Pinch your fingers closer and the diffraction improves.

We'll discuss diffraction again in Chapter 7 as it pertains to the interference of light, which creates many beautiful colors in both the natural and man-made worlds, including peacock feathers, butterflies, oil slicks, and bubbles. We will defer further examination of refraction in nature to Chapters 8 and 9.

5 Lenses: From Water Drops to Telescopes

5.1 Viewing the Unknown

On my office wall is an old photograph of a man peering through the end of a long tube. The other end of the tube is beyond the edge of the photo, pointing somewhere into the Great Unknown. The man is Percival Lowell, the tube is the great 24-inch refracting telescope at Lowell Observatory in Flagstaff, Arizona, and with it Lowell explored the boundaries of the outer solar system and the mysteries of the inner planets. This telescope and its brother, the 40-inch refractor at Yerkes Observatory in southern Wisconsin, stand as the pinnacle of astronomical

Fig. 5.1 Percival Lowell with the 24-inch refracting telescope of Lowell Observatory. Courtesy Lowell Observatory Archives.

technology of the late nineteenth century and the end of the era of lens-based telescopes. With the construction of the 40-inch, astronomers realized larger refracting telescopes were technically and economically unfeasible; construction of the 60-inch reflecting (i.e., mirror-based) telescope for Mt. Wilson (outside Los Angeles) began in 1904.

But why even use lenses to look into the heavens? We've seen how glass and water bend light. How can we use that simple concept to see things tiny and hidden as well as things far away and bright? This chapter explores the lens and its history, properties, and myriad uses in science and in everyday life.

5.2 The Focal Length

Most optical elements (e.g., mirrors, lenses) can be characterized by one quantity: the focal length. To know the focal length, we need to understand the focal point. The *focal point* is the point where an ideal lens converges light, or, in the case of a diverging lens, it's the point where the light seems to be coming from as it passes through the lens. Figure 5.2 demonstrates the focal point for a converging lens. Notice in the photo and schematic diagram that the light rays are parallel to each other and that they approach the lens from the left along the axis perpendicular to the lens (the optical axis).[1]

Our diagram models an ideal situation that is rarely found in nature. In reality, light rays are parallel only when they are artificially generated or when their source is very far away. Otherwise, light is divergent. The one major exception in nature is of course light from the Sun, which is so far away that the Earth is literally bathed in parallel rays of sunlight. In Fig. 5.3, a diverging lens refracts parallel light and spreads it out, diverging the light rays. In this case, the focal point is left of the lens, from where the diverging rays seem to come (a close approximation of the focal point is seen where the reflected rays intersect).

The focal point for a lens, then, can be at different places depending on where and how the light is impinging on the lens. For now, let's define the focal length within our ideal system, in which light shines on the lens parallel to the optical axis. Under such conditions, the distance from the lens to the focal point is the *focal length* of the lens. If the light encounters the lens from the right side, then there also would be a focal point on the left side of the lens, but the focal length is the same in either case.

During your next eye exam, asks the optometrist about your prescription. Let's say your prescription is –4.75. What are the *units* of that number? What does the negative sign mean? Optometrists use the arbitrary unit of *diopters* to

---

[1] Note also the multiple reflections from the exterior and interior surfaces of each lens. Light reflects *and* refracts at boundaries. Is the Law of Reflection obeyed?

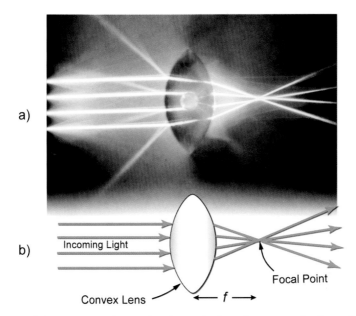

a)

b)   Incoming Light

Focal Point

Convex Lens   ← *f* →

Fig. 5.2(a) Light comes to a focus when it travels through a converging lens. (Photo by Steve Beeson) **(b)** A schematic of parallel rays of light converging at a focal point as they pass through a convex lens.

Fig. 5.3 Light spreads out when it travels through a diverging lens. Light is also reflected from the front and back surface of each lens in this figure and in Fig. 5.2(a). (Photo by Steve Beeson)

describe the focal length of the lens needed to correct vision. A diopter (which measures the power of a lens) is simply the inverse of the focal length, measured in meters. A positive or negative sign indicates the type of lens needed: A negative lens indicates a concave (or diverging) lens and corrects myopia, or nearsightedness. A positive lens indicates a *convex* (or converging) lens and corrects hyperopia, or farsightedness.

To measure the focal length of a lens yourself, find a distant light source (the Sun is ideal, but a high ceiling light will do) and use a converging lens to focus the light to a point on a piece of white paper. (Be careful imaging the Sun — you may burn the paper before your measurement is complete!) While holding the lens in place, measure the distance from the focal point to the lens. Its power in diopters is simply the measured focal length converted into meters, and that number is inverted. For example, for a measured focal length of 11.5 cm, the power in diopters is $D = 1 / 0.115$ m $= +8.7$. The positive sign reveals that it was a converging lens; diverging lenses are much harder to gauge, as we will see when we measure a lens' magnification.

5.3 Of Objects, Images, and Burning Glasses

Before we leap into measuring lenses, we should introduce a few more terms and ideas that will appear in this chapter. In our optical syntax, an *object* is simply anything that emits or reflects light. A lens bends the light and produces an *image* of the object. Anything can be an object — even an image! When creating multiple lens systems, scientists and engineers often propose that an image be the object of a lens, because an image is simply a place where light converges and the light continues to travel to the next lens, which then creates an image of the previous image (now the object). The eyepiece of a telescope or microscope, for instance, allows us to view the image of the image that the objective lens creates.

The image, then, is that area where light from an object converges or seems to converge. In the latter case, a concave lens diverges light that appears to come from the focal point on the objective side of the lens (the side where the object is). Convex lenses, on the other hand, converge light on the image side of the lens, as in Fig. 5.4(a). (We have used a cut-away view of the lens, as seen from the side.) The object side is to the left, and the image side is to the right, regardless of lens type. In Fig. 5.4(b), light from the candle passes through the convex lens and forms an image on a ground-glass screen.

A few more definitions now will allow us to make some comparisons between optical instruments a little later. *Resolution* is the smallest separation between two object points that a given lens (or mirror) can still show as two distinct entities. In practice it is the measurement of the smallest detail you can see with the lens. The *depth of focus* is the distance above and below the geometric

Object   Lens   Image

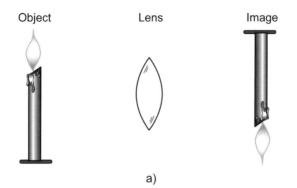

a)

Fig. 5.4(a) By convention, the left side of a lens is the object side, and the right side is the image side.

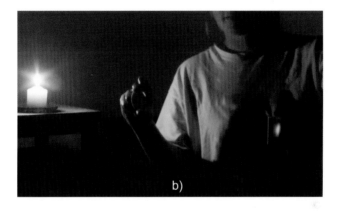

Fig. 5.4(b) A candle's image is formed on a glass screen that has been roughly polished. The rough glass scatters the light, allowing us to see the image. (Photo by Steve Beeson)

image plane within which the image is in focus, as in Fig. 5.5. You can test your depth of focus by closing one eye and bringing a hand-written note or a book closer to and farther from your open eye. With your eye relaxed, the range within which the letters and words are in focus is the depth of focus.

5.4 Burning Glasses

Imagine if you will that you are living in Europe in the sixteenth century. Clothes are heavy and uncomfortable, sanitation is poor if existent at all, and life is generally hard. To make matters worse, some "humors" have invaded your

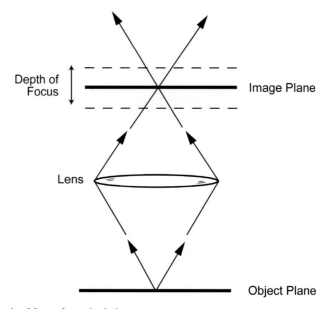

Fig. 5.5 Depth of focus for a single lens.

eyes, and your vision is blurred. One day you meet in the square a traveler from the Netherlands who lets you look through his "burning glass." Suddenly, the humors are gone! As you look through the glass, the world is clear, sharp, and, oddly, larger. How can the glass make things larger? What magicks are at work here?

The convex lens may not expel humors, but it does magnify things. Try it, dear reader: hold a lens to this page and look at the words. The letters are larger, but if you hold the lens away from your eye, the words look smaller, farther away, and upside down. What is going on?

When you hold the converging lens close to the book, the lens creates, on your retina, an image of the letter that is larger than the original letter (the object), as in Fig. 5.6(a). It does this by refracting the light toward the normal, just as we saw in Chapter 4. But since the surface of the glass is curved, the light rays that strike the bottom of the lens head off in a different direction than those at the top of the lens. They all head toward each other like a colony of ants zeroing in on a morsel of chocolate. They meet at the lens' focal point and continue their straight path as if the focal point were simply a mile marker on their journey. The image, then, is just a combination of multiple focal points, all meeting on the *focal plane* and recreating the pattern of light that was the object. Your brain processes the image, turns it right-side up, cancels out chromatic aberration, and places it in a three-dimensional space.

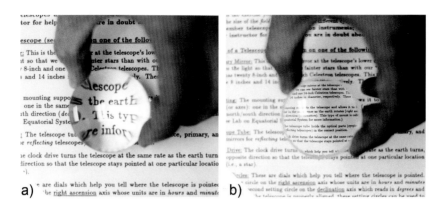

Fig. 5.6(a) A convex lens that is closer than the focal point to the object (the text) will magnify the object. **(b)** A concave lens always *minifies* objects. (Photos by Steve Beeson)

About concave (or *negative*) lenses and magnification, we will be brief. Concave lenses only minify, they do not magnify objects. The text of Fig. 5.6(b) is smaller when viewed through the concave lens placed anywhere in front of the page. These lenses create what Johannes Kepler, the great astronomer of the sixteenth century, called a *hanging image* — one that seems to hang in space between the lens and the object. We call it a *virtual image* because it can be seen, but it cannot be projected onto a screen. This is why you may have difficulty finding the focal length of a concave lens — there is no image on the paper to measure!

Convex (or *positive*) lenses can project images on screens. These lenses are used in film and slide projectors and were known as *burning glasses* in the Middle Ages. Like concave lenses, though, convex lenses can also produce virtual images. The lens will produce a *virtual image* if the lens is close to your eye, and it will produce a *real* image if the lens is far from your eye.

5.5 Measuring Magnification

Magnification is simply the measure of how large an image is compared to its object. In the language of mathematics, lens magnification is the ratio of the image size to the object size. Usually, we compare the *height* of the object to the image, but most lenses are made to magnify proportionately, so in theory, any dimension could be measured and compared. After some handy geometric manipulation, we find that the magnification of a lens is simply the negative of the ratio of the image distance and the object distance. That is,

Fig. 5.7 A Leeuwenhoek microscope. The scientist would place the specimen on the needle, using the screw to adjust the needle's height. The lens hole is in the plate, to the right.

Magnification = image length/object length =

- image distance/object distance

The negative sign tells us whether the image is inverted or upright. Look through your lens at the page again, this time with your eye relaxed and far from the lens. The words should appear right-side up and larger — this is a virtual image. You can measure the magnification by placing a ruler next to your lens as you view a letter on a page (or a drawn arrow) and measuring the height of both the object and image. Most magnifying glasses or lenses magnify about 2.5x (M=2.5).

Some single lenses have much higher magnification. Antony von Leeuwenhoek, a Dutch scientist of the seventeenth century, achieved magnifications of up to 250x with tiny spherical lenses he manufactured for his friends and colleagues. One of these is seen in Fig. 5.7. The Leeuwenhoek microscope is simplicity itself. It has a single lens mounted on a metal plate, with screws to move the specimen across the field of view and to focus its image. The lens permits 70x to 270x magnification. Using these extremely precise microscopes, he discovered the world inside a smear of blood and the life teeming in a droplet of canal water. Three hundred years later, our best optical microscopes are only slightly better than von Leeuwenhoek's little masterpieces.

Leeuwenhoek microscopes contain a tiny, spherical lens made of glass. A drop of water on a clear plastic sheet can be used to make a similar lens, one that is plano-convex (one surface flat, one surface curved out). A water-drop magnifier can be made with a hole punched in a stiff card. A piece of clear plastic is placed over the hole, and a drop of water is put on the plastic over the hole to makes a water-Leeuwenhoek microscope, as in Fig. 5.8.

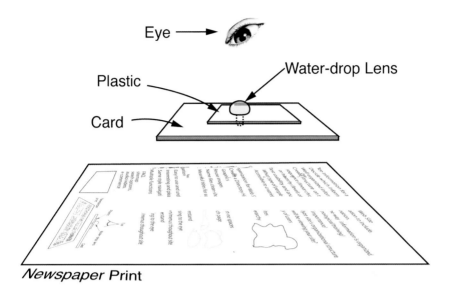

Fig. 5.8 Schematic of a water drop magnifier.

5.6 The Compound Microscope

Not everyone wants to look through a microscope no larger than a pinhead. Robert Hooke, who couldn't bear the eyestrain of the single lens and who introduced us to waves in the last chapter, also dabbled in microscopy. Years earlier than von Leeuwenhoek, Hooke built an inverted telescope using two lenses (based, supposedly, on a design by Galileo) and saw things no one had seen before — molds, tiny insects, and "cells," the compartments in plants used to store nutrients and water. His microscope, now called a *compound microscope*, consisted of a lens to gather and bend light from an object (the objective lens), and another, more powerful lens near the eye (the eyepiece) to magnify the image created by the first lens. The simple compound lens system seen in Fig. 5.9 is still the basic design used in laboratory microscopes and refracting telescopes. It allows for 1000x magnification, which makes it up to four times more powerful than the simple microscope. The objective lens system can be quite complex, using doublet lenses (a combination of two lenses of different materials) to correct *chromatic aberration* — the spread of an image over a range of colors.

With compound microscopes, the image from the eyepiece can be focused on an array of light-sensitive semiconductor devices, known as charge-coupled

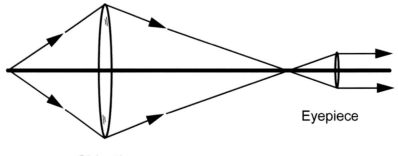

Eyepiece

Objective

Fig. 5.9 A simple compound lens. The eyepiece is a short focal-length converging lens. The objective has a longer focal length.

Fig. 5.10 The Drum Optical Microscope.

devices, or CCDs. These CCDs convert light into electrons, and thus they generate electronic images. Such images can appear on a TV monitor and can be stored on magnetic or digital media. Each light-recording element of a CCD is called a *pixel* (short for *picture element*), a term that has become popular because of digital cameras and their megapixel image sizes. (A megapixel is one million pixels.) Astronomers use CCDs to capture their heavenly images in both ground-based telescopes and space-based instruments, such as the Hubble Space Telescope, whose Wide Field Camera 3 (WFC3) instrument has 8.4 megapixels in each of its two CCDs.

Figure 5.10 shows a Drum microscope made in the 1800s, the midpoint of the development of compound microscopes that began in the very early 1600s and carries on today. The sample would be placed on the glass plate at the bottom of the microscope, with the objective lens just above it. The objective is removable to allow different magnification depending on the focal length of the lens inserted. Near the top of the microscope is the eyepiece; the knob in the middle allowed the user to adjust the objective/sample distance and thus focus the instrument. There are hundreds of variations of these elegant microscopes. For more information, see Gerard Turner's *The Great Age of the Microscope* (1989).

5.7 A New Microscope

Besides the simple beauty of a good microscope, the attractive feature of optical microscopy is that it is so easy: samples can be analyzed in air or water; the images are in natural color with magnifications of up to 100,000x; and modern semiconductor electronics with CCDs allow fast image processing. The optical microscope should dominate the field of microscopy. It doesn't. Let's see why.

Figure 5.11 shows slices of cork drawn by Robert Hooke in 1665. The image was taken from his book *Micrographia*, which introduced the macro world to the micro world of cells, lice and stinging nettles seen through the new compound microscope he invented. Figure 5.12 shows a similar piece of cork observed through a Scanning Electron Microscope (SEM) and magnified 400x. The SEM extends the limited magnification range of the optical microscope — which normally extends to only about 1500x — to more than 50,000x. It's especially useful when analyzing the images of specimens that have a great deal of

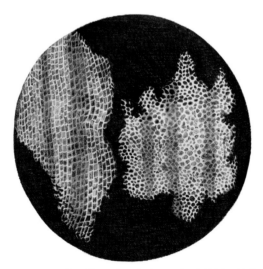

Fig. 5.11 An optical image of a slice of cork drawn in 1665 by Robert Hooke. Taken from *Micrographia*.

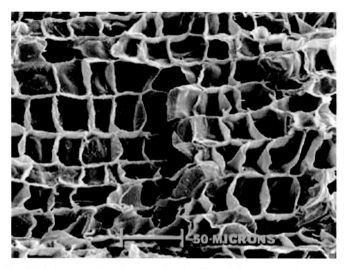

Fig. 5.12 SEM image of cork magnified 400x. Taken from Images of Nature, http://ion.eas.asu.edu

surface relief, i.e., that have features such as cracks, fibers, indentations, and rough surfaces.

However, the two key advantages of the SEM are its incredible depth of focus and its *resolving power*. We noted earlier that the depth of focus is the range in distance over which a specimen is in focus. For an optical microscope, this range is typically only a few percent of the linear size of the image. The SEM, on the other hand, extends its depth of focus over several percent of the image size. Note the differences between the images of a tungsten filament (which lights household light bulbs) in Figs. 5.13 and 5.14. Figure 5.14 is also a good example of the SEM's superior *resolving power* (the smallest detail that a microscope can resolve, or see). A good optical microscope under ideal conditions has a limiting resolving power of about 0.2 microns (200 nanometers, nm). This means that if two objects on the specimen are closer than 0.2 microns, the optical microscope will not be able to resolve them.

Thinking of it another way, if you're driving on a long, straight road at night and you see a car coming toward you far in the distance, you may not be able to see, or resolve, two headlights. Only when the car is closer will you be able to see both headlights. In that scenario, the resolving power depends on your eyes and the amount of light scattering in the air. The resolving power of the microscope depends on the optics and lenses inside and the wavelength, or type, of light used to see the specimen. For scanning electron microscopes, the resolving power is orders of magnitude (powers of 10) better than those of optical micro scopes because the wavelength of light used is orders of magnitude smaller.

Fig. 5.13 Optical image of tungsten filament. Taken from Images of Nature, http://ion.eas.asu.edu

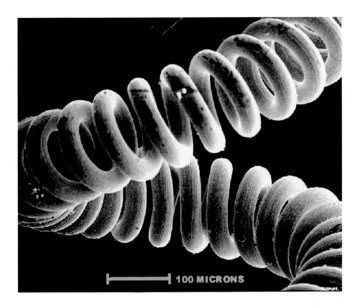

Fig. 5.14 SEM image of a tungsten filament. Taken from Images of Nature, http://ion.eas.asu.edu

5.8 Inside the SEM

How does the SEM use light to obtain such highly resolved images? Actually, it doesn't use light at all. As its name suggests, the Scanning Electron Microscope probes the unseen world with a beam of electrons, not a shaft of light. These electrons have wavelengths much smaller than the wavelengths of light used in optical microscopes, and this allows smaller details to be seen.

Imagine that you're trying to determine the layout of a rough surface that you cannot see. You could drop a handful of balls on the surface and measure how they bounce. Using smaller balls would improve your chances of determining the true layout, because small balls would bounce off small details. A BB pellet would work better than a golf ball, which itself would work better than a beach ball. Following the analogy, the SEM electrons are the BB pellets, and visible light is the beach ball. The smaller the wavelength (or size) of the probing unit, the greater detail it can see.

Although the SEM uses electrons, it doesn't use lenses like a standard optical microscope does — at least not in the traditional sense. The lenses in a microscope or telescope are made of glass, which refract and focus light. Since the SEM's probing unit is the electron and not light, one must move and focus the

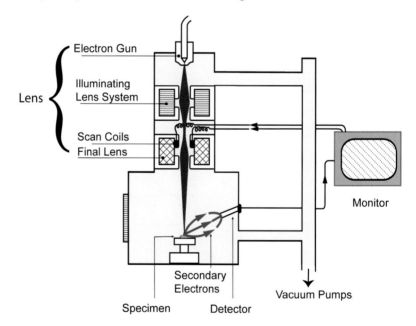

Fig. 5.15 Schematic of a Scanning Electron Microscope.

electrons using something that the electrons respond to: electromagnetic fields. Small coils of wire generate magnetic fields through which the electrons pass on their way to the specimen. Functions such as brightness, focus and magnification can be fine tuned by altering the levels of electric current in the coils. Currents must be very precise, and they must remain stable once set. A cut-away diagram of a SEM is shown in Fig. 5.15, which describes three main parts of the microscope: the column, the vacuum system, and the detector and monitor.

At the top of the SEM column is the chamber, housing the source of the electrons (also called the *electron gun*) and a heated tungsten wire with its associated electrodes. The power supply provides the high voltage necessary to accelerate the electrons towards the specimen. The voltage determines the wavelength of the electrons in the beam, and in a conventional SEM this requires a stable voltage of between 500 and 40,000 volts. Below the electron gun is the lens system, consisting of a series of electromagnetic lenses stacked on top of one another. The coil of wire that forms each lens is shielded from external magnetic fields and cooled by refrigerated water.

The magnetic lenses focus and move the beam across the sample, which ejects secondary electrons when struck by the SEM's electron beam. These secondary electrons are like the BB pellets bouncing off the rough surface; if we measure the energy and direction of the pellets, we will have a view of the layout of the surface. The electron beam moves across an area known as the *raster*, in which the sample is placed, as in Fig. 5.16. As the beam moves across the

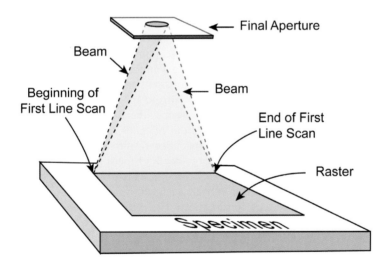

Fig. 5.16 Image formation by scans of the electron beam on the specimen.

line, an electron detector measures the strength of the secondary electron emission from the surface. The microscope repeats the scan one line down again and again until the whole raster area is scanned, which takes a thirtieth of a second.

Typical SEMs use 1000 line scans to form a 10 cm x 10 cm image. A television monitor displays the signal from the detector — a line-by-line electron scan of the sample.

A very important part of the SEM system is the system of vacuum pumps that remove air from the space inside the column. A good vacuum is necessary to allow the electrons to move unimpeded down the column (which may be a meter in length from the electron gun to the specimen). Electrons will not travel very far through air before being absorbed. The vacuum system also minimizes contamination of the specimen resulting from interactions between the electron beam and residual gas molecules.

5.9 Optical versus Electron Microscopy

The attractive feature of optical microscopy is that it is so easy. Samples can be analyzed in air or water, the images are in natural color with magnifications of up to 100x to 1000x, and modern semiconductor electronics with charge-coupled devices (CCDs) allow the image to be processed by increasingly sophisticated computers and software. The optical microscope should dominate the field of microscopy. It does not. The scanning electron microscope (SEM) is the microscope of choice because of its depth of focus and resolving capability, as seen by Figs. 5.13 and 5.14, which demonstrate the striking differences in the images of a coil of tungsten filament. In the optical microscope photograph (a *micrograph*) taken at high resolution, only sections of the filament are in sharp focus, while in Fig. 5.14 the whole specimen is in focus.

For the optical microscope, the depth of focus indicates the distance above and below the image plane over which the image appears in focus. As the magnification increases in the optical microscope, the depth of focus decreases. (Note the distant part of the filament is out of focus.) In the SEM image, the three-dimensional appearance of the specimen results directly from the large depth of focus of the SEM. It is this large depth of focus that is the most attractive feature of the scanning electron microscope. This feature arises because the data is obtained with a fine electron beam scanned over the surface and with the detected secondary electrons forming an image on the TV-like monitor. The SEM sees the entire specimen in focus because it probes the entire specimen bit-by-bit, whereas the optical microscope sees the entire specimen all at once and so can only have a narrow range of focus. Furthermore, as noted earlier, the resolving power of an electron microscope is far superior to that of an optical microscope because the wavelength of the probing beam (UV electrons versus visible light) is orders of magnitude smaller. The disadvantages of the SEM are

Fig. 5.17 Optical microscope image of Alyssum. Taken from Images of Nature, http://ion/eas.asu.edu

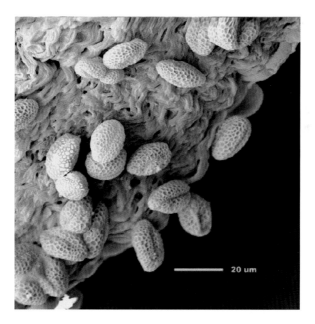

Fig. 5.18 Scanning Electron Microscope image of the pollen in an Alyssum at a magnification of about 1000x. Taken from Images of Nature, http://ion.eas.asu.edu

numerous, though. Since electrons are lightweight (1/1836 the mass of the proton) and scattered or absorbed in air, the sample chamber must be in vacuum that limits the sample size to a few centimeters. Electrons carry charge, so the samples must be covered with a conducting coating — often gold plating — which can be expensive if the SEM is used often. Finally, the SEM itself is expensive and requires regular maintenance. Unlike the colorful optical micrograph seen in Fig. 5.17, the SEM generates black-and-white images (Fig. 5.18), although at a much greater magnification and resolution than the optical microscope. Both images show the hidden beauty of the natural world.

5.10 Seeing the Distant

Percival Lowell sat at the 24-inch telescope atop Mars Hill in Flagstaff, Arizona, night after night, sketching the dark valleys and bright plains of planet Mars. Through the great 24-inch, Lowell believed he saw thousands of canals, gardens, and earthworks built by Martians. Of course we know now that his imagination got the better of him while he sat in the cool, northern Arizona night. Yet he and his telescope started a small cultural revolution. At the turn of the century, his writings inspired millions to speculate on life outside our blue planet. New novels, plays, and music were written celebrating the red planet and its mysteries.

No less a sensation gripped sixteenth-century Europe after Galileo built one of the first telescopes and spied four major moons orbiting Jupiter. Refuting the traditional Earth-centric view of the universe, Galileo's discovery supported Copernicus' idea that all the planets orbited the Sun, like Jupiter's moons did their king-sized master. These views did not sit well with the Roman Catholic Church, and Galileo's continued, vocal denunciation of the Platonic (Earth-centered) cosmology eventually landed him under house arrest and near excommunication. In the end, Galileo hastened in the scientific age with his telescope in hand and his radical ideas of using experiment rather than theory alone to verify natural law.

At the most basic level, telescopes and microscopes serve the same purpose: they magnify the tiny. For the microscope, the tiny truly are tiny — insects, microorganisms, cells, viruses, and even individual atoms. For the telescope, the tiny are large objects that are far, far removed. Since telescopes and microscopes do basically the same thing, it follows that they are built in similar fashion. The main difference is that, to capture the very dim light from distant stars and galaxies, telescopes must be much larger. To compare the relative sizes of the two instruments, consider that most optical microscopes have an objective lens smaller than an inch in diameter.

Two of the largest telescopes in the world are the Yerkes 40-inch refractor in Williams Bay, Wisconsin, and the Very Large Telescope (the VLT). The Yerkes is the largest lens-based telescope in the world. The VLT has four 8-meter reflecting (mirror-based) telescopes in combination — essentially a 50-foot (600-inch) diameter mirror. These telescopes allow astronomers to see objects that are exceedingly far away and, because the universe has expanded over time, exceedingly old. Working in its dual-mirror mode, the VLT can see objects that were created after less than 1 percent of the current age of the universe had elapsed.

Unlike the simplest microscopes, refracting telescopes use two lenses. Like the compound microscope, though, the names of the optical elements (lenses or mirrors) are the same: the objective lens is closest to the object, and the eyepiece is near the eye. The simplest type is simply that — a tube and two converging lenses. And actually, you don't even need the tube. The objective lens creates an image of the distant object, and the eyepiece (a magnifying glass, essentially) enlarges that image. In all types of telescopes, changing the eyepiece changes the magnification of the image — the smaller the focal length of the eyepiece, the greater the magnification. Galileo's telescope used a diverging lens as its eyepiece, a practice uncommon now except in telephoto lenses and opera glasses.

Telescopes also come in the reflecting variety. These are the types most often used for astronomical research and serious amateur observing. Instead of an objective lens to capture starlight, reflectors use large mirrors at the back of the telescope tube to reflect and focus the light through a series of more mirrors and lenses and finally to the eyepiece. Isaac Newton invented his own brand of telescope by placing the eyepiece at the side of the tube, making observing a bit more comfortable than getting behind, and sometimes under, the telescope. There are many types of telescopes available, so take advantage of the many stargazing opportunities each year to observe the telescope varieties of your local astronomy club.

5.11 Imperfect Light

In May of 1990, astronomers at the Space Telescope Institute in Baltimore received from the new Hubble Space Telescope the first images of NGC 3532, a sleepy star cluster in the constellation *Carina*. This particular cluster was special simply because the images from it would tell astronomers if the mirrors, lenses, and sophisticated electronics on board the space-based telescope had survived the launch and deployment and the harshness of space. Within hours of first light, the Space Telescope team knew something was wrong: the images downloaded from the telescope were blurry and strangely shaped. A few days later, the engineers and scientists had solved the puzzle. The space telescope contained a flaw — an aberration — in the primary mirror, which is the major light-gathering element. That aberration ended up costing several hundred million dollars to fix in a service mission to the telescope three years later (Chaisson 1998).

The Hubble Space Telescope story is an extreme tale of optical systems gone wrong, but anyone who builds a precise optical instrument must deal with issues like these. Every spherical lens or mirror exhibits *spherical aberration* because light from different parts of the lens (or mirror) gets focused to a different place. Ideal lenses focus all light incident on them to a single, sharp point. Light that comes from the edges of real lenses, however, comes to focus before those near the center of the lens. This creates a blurry edge, a *circle of confusion*, around the image. The aberration can be reduced by masking off the edge of the element (stopping it down) or by using a smaller lens or mirror. Even these measures don't entirely eliminate the problem, however. A lens shaped like a paraboloid or hyperboloid will correct spherical aberration as well as a compound lens system will.

Our eyes have lens aberrations, even beyond the common near- or farsightedness. Most of the aberrations of the eye are corrected in one way or another, but one, *chromatic aberration*, still plagues our vision. As we will discuss in the next chapter, the change of refractive index with wavelength is called *dispersion*, and it's responsible for rainbows and the "fire" in diamonds. In the simplest terms, in a dispersive medium such as glass, blue light is bent more than green light, which is bent more than red light. All lenses, then, will bring to a focus the blue light from a light source before the red, and images will have a circle of confusion that looks like a rainbow. Figure 5.19 shows a side-view diagram of a lens and its chromatic aberration.

This chromatic aberration (from the Latin *chroma*, for color) arises mostly because the light passes through the lens' edges, which are like little prisms.

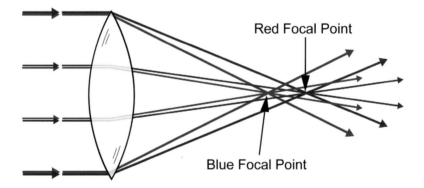

Fig. 5.19 All single lenses exhibit some chromatic aberration, in which light of different colors is focused in different places.

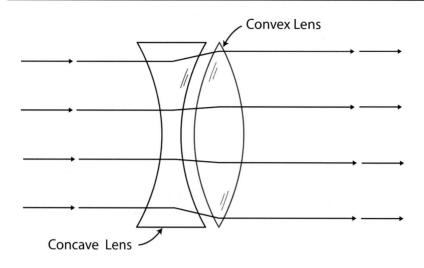

Fig. 5.20 An achromat doublet — a convex and a concave lens put together.

One way to correct it is, again, to stop down the lens. To avoid light loss, however, most people prefer to correct the *aberration* by putting two lenses of different refractive indices next to each other; this is known as an *achromat doublet*, a technique hypothesized (and thrown out) by Newton but re-invented by a British lawyer two years after Newton's death. Most doublets consist of a convex and a concave lens, as in Fig. 5.20. Our eyes do not contain an achromat doublet (though we do stop down our lenses with pupils), but thankfully the brain adjusts to chromatic aberration when it processes the visual signal.

There are several other aberrations that lenses suffer, such as astigmatism, coma, field curvature, and distortion, but we will leave these details to other volumes.

5.12 The Most Advanced Camera

No sophisticated optical system or combination of lenses is as complex or as fascinating as the mammalian eye. Not content with just self correcting for aberrations, the eye can quickly alternate its focus, and it can easily adapt to low- or high-level light situations. With its partner, the brain, it can distinguish between shades of color with a finesse still unmatched by technology.

Like most of the body's organs, the eye's parts, illustrated in Fig. 5.21, work together wonderfully. The eye's first job is to refract light. This task falls to the *cornea*, the outer, hemispherical button at the front of the eye. Enclosing the liquid-filled *aqueous humor*, the cornea has a refractive index of 1.376, higher than water or most glass. Light enters the eye and is bent at the air-cornea interface. Covering the cornea is a thin layer of protective skin called conjunctiva, which is primarily known for becoming irritated and infected, leading to conjunctivitis, or pinkeye.

Notwithstanding disease, the light next passes through the *pupil*, the small aperture in the *iris*. The pupil loosely controls the amount of light entering the eye. When we look at our eyes in the mirror, the pupils look black because we're looking inside each eyeball, where no light is emitted. If we could see inside, we would see the faint outline of the *crystalline lens* and a sea of red — the *retina*. This sea is, in fact, what you see when "red eye" plagues the photos of your family reunion: light from the flash is reflecting from the blood-infused layer of retinal cells in the eye's interior.

Like a tiny pearl onion being pushed and pulled, the crystalline lens has a thankless job. Not only is it a secondary refractive device in the eye (the primary distinction goes to the cornea), but it can be removed altogether without profoundly affecting vision! Formed throughout our lives by adding successive, onion-like layers of tissue, the eye lens' primary task is to refine the focus of incoming light. When relaxed, the lens is relatively thin, allowing light from distant objects to slightly bend as it passes through. When *accommodating*, the muscles surrounding the eye contract and compress the lens to make it thicker. Nearby objects, such as the words on this page, reflect or emit light that is sharply divergent when it reaches the eye. To accommodate for this non-parallel light, the lens becomes fatter and further refracts the light received from the cornea. This is why you can sometimes look closely at an image and then still see it when you close your eyes. The strain of the eye muscles compressing the lens often causes headaches after hours of reading or computer use.

The lens can also safeguard against unwanted high-energy ultraviolet light that can enter the eye and damage the retina. However in blocking UV light, the lens often performs its job too well and blocks some visible blue light. As one gets older, the thin layers of the lens tend to harden and yellow, blocking even more blue light and losing the ability to finely accommodate nearby objects. With extreme age, most lenses develop cataracts — deposits of dead cells that blur the image and inhibit vision. For this reason, lenses with severe cataracts are often removed and, because the lens is the UV blocker, such old eyes can often then see many ultraviolet sources of light.

After passing through (or not) the crystalline lens, light from an object crosses the *vitreous humor*, filling the inner eye and striking the retina. There, light is converted into chemical and electrical signals that are sent to the brain. We will discuss the retina and its crucial role in vision and the perception of color in the next chapter.

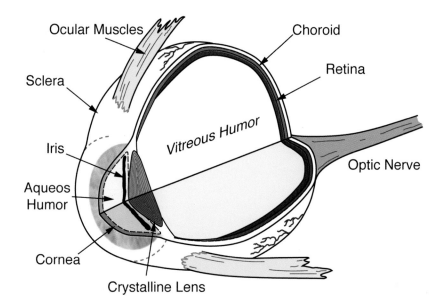

Fig. 5.21 A cut-away diagram of the human eye. The crystalline lens is like a bean-shaped pearl onion that can flex to become thicker or thinner, depending on where the incoming light is focused.

5.13 Still an Imperfect Camera

As precise and beautiful as the eye is, it is not without its flaws. Cataracts plague the elderly, glaucoma strikes the unsuspecting, and pinkeye haunts the unsanitary. More commonly, a large number of the world's population suffers from myopia, hyperopia, or presbyopia.

The nearsighted, myopic eye is too long — the eye is too powerful. Light from distant objects comes to focus before the retina, so the objects are difficult to image. The divergent light of closer objects is a blessing to the myopic, as he can often see those objects with fine detail. A diverging lens corrects the problem. The focal length of this lens is equal to the *far point* of the relaxed eye, or the farthest distance from which an object can be clearly seen. The shorter the far point, the stronger the prescription. Returning to our unit of diopters, which is the inverse of the focal length (in meters), a prescription of –6.75 diopters tells us two things: First, that the lens will be diverging (the negative sign gives it away), and second, that the focal length of a corrective lens will be $1/6.75 =$

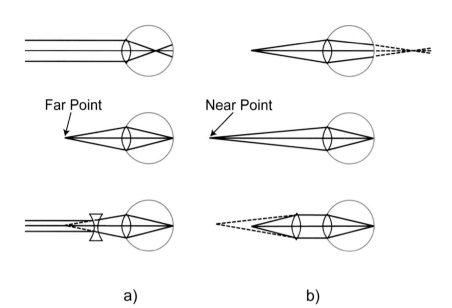

Fig. 5.22(a) The far point of the myopic eye is closer than the far point of an average eye. A diverging lens is needed to bring distant objects to comfortable focus. **(b)** The near point of the hyperopic eye is farther than the near point of an average eye. A converging lens is needed to bring close objects to comfortable focus.

0.148 m = 14.8 cm. That's a pretty strong lens. And a pretty strong eye. The lens power indicates that the farthest that an object can be seen clearly is only 15 cm. (For an average eye, that distance is several meters.)

If reading the deli sign across the street poses no problems but reading a book has become difficult, you may be suffering from *hyperopia*. In this case, the eye is too short and objects close by are out of focus because light entering the eye focuses beyond the retina. A farsighted person can usually see distant objects without trouble, but the *near point*, the closest that an object can be while staying in focus, is farther than that of an average eye. A converging lens corrects for hyperopia.

An unassisted farsighted person would have trouble focusing on an object closer than this point, especially to see, for example, the small print of a book. Many people believe they are suffering from hyperopia when, in fact, their lenses are just old and tired. *Presbyopia* is the loss of the crystalline lens' ability to accommodate, or focus, and requires corrective lenses for short distances. And if you're myopic as well, it's bifocal time!

Luckily, advantages in lens-making technology have allowed even severe my-, hyper-, and presbyopes to enjoy lightweight but powerful eyeglasses and comfortable contact lenses. Another option is an adjustment to the eyeball itself: refractive surgery uses a laser to adjust the shape of the cornea, and it has become a very popular (if somewhat expensive) treatment of myopia.

Of the large number of eye maladies, perhaps the most troubling is color deficiency, or *color blindness*. Not truly blind to color, the color deficient has usually lost some color sensitivity in the retinal cells and hence cannot easily discriminate between certain colors. Color deficiency occurs primarily in males and is usually manifested in the failure to differentiate between many red and green wavelengths. Appropriately, with the profusion of color computer monitors and access to the Internet in the workplace, steps are being taken to provide "color styles" to Web sites so that color deficient users may more easily distinguish between the millions of colors displayed on their computers.

The role of color in vision is a complex one. Although most people can come to a common agreement on what "blue" is, each of our visual systems are different: two people may, for instance, have very different opinions on the color of a blue-green (or green-blue) dress. In the next chapter, we will explore color — its composition, sources, and mystery — and more closely examine how exactly it is we see color.

6 Sources of Light and Color

6.1 Crossroads

In Chapter 4 we examined the properties of waves and how they explain the long-unexplainable phenomena of refraction (why light bends) and diffraction (why objects cast fuzzy shadows). But in the eighteenth century, Newton's followers still controlled the optical ideas of the time with their theory of corpuscles. In the end, Newton's experiments in diffraction and color theory undermined his own corpuscular theory and ushered in the great era of electromagnetic unification led by Faraday, Hertz and James Clerk Maxwell. But what sublime experiments Newton imagined.

6.2 Waves, Rebounding

Just after receiving his bachelor's degree from Cambridge, Isaac Newton retreated to his mother's house in Lincolnshire, England, during the plague scare of 1665–1667. When not creating the calculus and pondering gravity, he kept busy by revolutionizing optics and color theory. Equipped with two prisms and some sun, Newton changed our views of light in general and of sunlight in particular. He determined that light is not fundamentally changed by a prism, as had been thought, but is simply bent by it. The amount of bending, however, was determined by the multitude of wavelengths that compose light. He found that blue light was bent the most, and red was bent the least. Furthermore, the full range of colors in sunlight brought out by the prism, the *spectrum* of visible light, was presented in a precise array from red to orange, yellow, green, blue, indigo, and into violet. This is illustrated in Fig. 6.2 and photographed in Fig. 6.1, in which a hand-held prism produces the colorful array from a shaft of sunlight.

Newton said the order of colors was constant, and each color had a unique signature identifying its location in the spectrum. Though he resisted the idea of light as waves, his great gift to color theory was the concept that a light's wave

Fig. 6.1 The dispersion of white light is shown by the spectrum of light passing through a prism. (Photo by Steve Beeson)

length is the signature of its color. He dismissed the wave theory of light that was proposed by Hooke and Huygens (as we saw in Chapter 4) but suspected color had something to do with waves when he wrote in Optiks (Query 14):

> May not the harmony and discord of Colours arise from the properties of the Vibrations propagated through the Fibres of the optick Nerves into the brain, as the harmony and discord of Sounds arise from the properties of the Vibrations of the Air?

In the century after Newton, progress in optics and color theory fell behind electricity and the newly rejuvenated field of chemistry. The *emissionist* theory of light (Newton's corpuscular theory, evolved a bit) held sway despite growing evidence against it — mainly from the experiments of Young and the calculations of Fresnel. Throughout the nineteenth century, scientists slowly turned to the wave theory as Newton's complaints about it fell one by one to the hard realities of laboratory evidence.

A century and a half later, at the end of the American Civil War, a Scottish physicist named James Clerk Maxwell took a hard look at the various equations

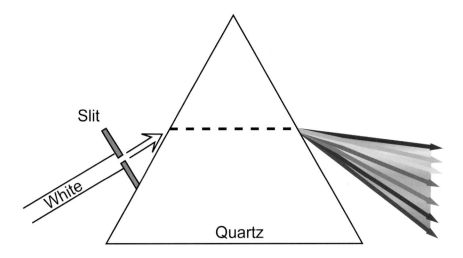

Fig. 6.2 White light directed through a prism is refracted, with dispersion bending each color differently. Violet has the greatest dispersion.

of electricity, magnetism and waves that had been developed since Newton's *Opticks*. He concluded that light is simply a form of *electromagnetic radiation* traveling as a wave at a constant speed. The speed of the radiation is the speed of light. In one blow, Maxwell unified the three seemingly unrelated concepts of electricity, magnetism, and light into one grand theory that perfectly predicts radio, microwaves, and television. His work also advanced the field of astrophysics, in which scientists equipped only with telescopes and the basic laws of physics observe and attempt to explain the greater universe. It is the theory upon which our current technological infrastructure is built. Of course it took the scientific community twenty years to accept the theory because it proposed so much, including the fact that visible light, upon which much of our sensory information is based, is only a fraction of the electromagnetic spectrum of energy; Maxwell's electromagnetic theory predicted light exists beyond the red and the violet.

Fifty years before Maxwell's theory, radiation had been discovered in just these extra-visible regimes, but the proper explanation of *infrared* and *ultraviolet* light demanded a complete energy spectrum, which was discovered piece by piece throughout the entire nineteenth century. Paul Villard discovered the last piece, gamma rays, in 1900, just as a young Albert Einstein was finishing his undergraduate work. Only a few years later would that young German scientist upend the prevailing paradigm and reestablish the credibility of the photon.

6.3 Waves, Unfolding

Figure 6.3 shows the electromagnetic spectrum extending beyond the visible radiation we call light to include, at one end, radio waves and at the other, gamma rays. The visible light region occupies only a very small portion of the spectrum, but it is this tiny piece to which our eyes are sensitive, mainly because the Sun emits most of its radiation in this band. We will examine the Sun and other sources of light later in this chapter.

Returning to Newton for a moment, we see in Fig. 6.4 the spectral sequence of colors as a circle — a color wheel — as he first envisioned after assigning names to the colors he saw projected on his parlor wall. The students' mnemonic is "Roy G. Biv," with each letter representing the first letter of each successive color: red, orange, yellow, green, blue, indigo, and violet. Newton noticed that purple (which is not in the spectrum) was as similar to his perception of red as it was to violet, and he began to wonder, Why should the spectrum be linear? He regarded purple as a mixture of red and violet, gave it its own slice of pie, and closed the spectral loop.

When we think of light as an electromagnetic wave, we can identify a color's place within the spectrum — its spectral signature — by measuring its wavelength. We sense the waves as color, violet being the shortest wavelength, near 400 nm, and red the longest, near 700 nm. Three typical waves of visible light are shown in Fig. 6.5. The wavelength is represented by the Greek letter λ and represents the distance between successive wave crests, often measured in *nanometers* (nm), a unit of distance in the metric scale. One nanometer equals one thousand millionths of a meter (m) or 1 nm = 10^{-9} m, or about 0.00000025 inches. One hundred wavelengths of visible light are smaller than the thickness

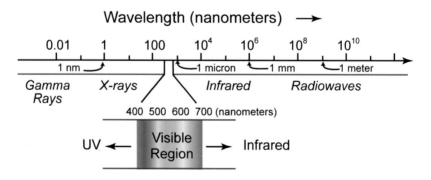

Fig. 6.3 The electromagnetic spectrum, which encompasses the visible region of light, extends from gamma rays with wavelengths of one hundred billionth of a meter to radio waves with wavelengths of one meter or greater.

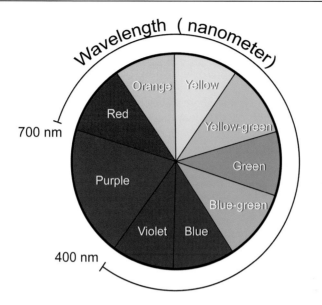

Fig. 6.4 The region of visible light in wavelengths shown as a circle was conceived by Isaac Newton. The purple region shown in the color circle is a mixture of light in the red and violet regions of the spectrum. Purple cannot be represented by a single wavelength (energy) of light.

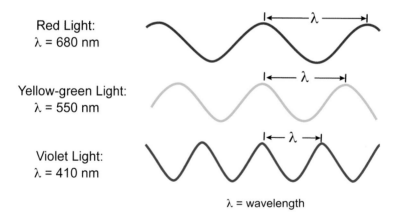

Fig. 6.5 A wave representation of three different colors: red, yellow-green, and violet, each with a different wavelength, λ, which represents the distance between successive wave crests.

of a sheet of paper or the diameter of a human hair.

As we travel through the visible spectrum of violet, blue, green, yellow, orange, and red, the wavelengths become longer. Figure 6.3 shows how the visible region (400–700 nm) is centrally located in the electromagnetic spectrum. Infrared (IR) and radio waves are at the long wavelength side while ultraviolet (UV), x-rays, and gamma rays are short wavelength radiation. The spectral regimes beyond the visible are, by definition, invisible to the human eye, but not to all animals. Many insects utilize portions of the UV spectrum to locate pollen-rich flowers, and snakes can focus the IR radiation emitted by warm-blooded prey.

6.4 Photons, Reflecting

As scientists struggled to make sense of the evidence that electromagnetic waves traveled through the universe without traveling *in* something (such as water or air), Einstein pondered a simple experiment in which light shined on a piece of metal. Pretty innocuous stuff, until he realized that the metal ejected *electrons*, the tiny elemental particles discovered only seven years earlier by the English scientist J.J. Thomson. How could waves of light so profoundly affect particles of matter? It would be like calm water waves pushing a large rock off the beach!

Einstein proposed that the light bulb emitted electromagnetic radiation not in the form of waves, but in tiny packets of energy he called *photons*. Einstein said that these photons, basically particles of light, carry information about the energy, and hence the wavelength, of the light wave. A single photon of one color (or wavelength in our wave analogy) differs from another photon only by its energy. In this description of light, the most convenient unit of energy to use is the *electron volt*, eV, which is the energy gained by an electron moving across a positive voltage of one volt. This is a very small amount of energy, but it's large for the tiny electron.

In this scheme, light is composed of photons that can have a range of energy from tiny fractions of an electron volt in radio waves to thousands of electron volts for gamma rays. Figure 6.6 returns to the spectrum of Fig. 6.3, but with the schema of photon energy alongside the light wavelength.

Visible light rays are composed of photons in the energy range of about 2 to 3 eV. As the energy of the light increases, the wavelength decreases — long wavelength light has low energy, short wavelength light has high energy. Orange light, with a wavelength of 620 nm, is composed of photons with energy of 2 eV. It is the photons with energies between 1.8 and 3.1 eV (or wavelengths between 700 and 400 nm) that trigger the photoreceptors in our eyes. Lower energies (longer wavelengths) cannot be detected by the human eye, but can be detected by some animals and special infrared sensors. Higher energies (shorter wavelengths) such as x-rays are detected by x-ray-sensitive photographic film or by special detectors.

We have discussed light in the particle sense and in the wave sense. In reality, we can think of it as either. A red beam of laser light can have a single wavelength of 650 nm. Equally, we can say the laser has an energy of 1.7 eV. However, it is useful to think of light as one form or the other when talking about certain parts of the spectrum. For example, radio waves seem to be more intuitive than radio particles or rays. On the other hand, gamma rays or x-rays make more sense than "x-waves." This is just a simple reference to the energy of the light. At the highest energies, it seems easier to think of light as a particle: singular, penetrating, exact. At the lowest energies, it is easier to think of it as a wave: long, undulating, and pervasive.

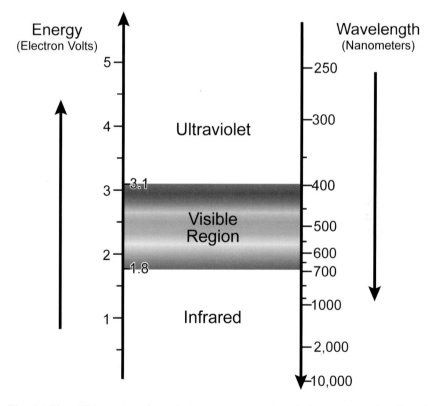

Fig. 6.6 The visible region of the electromagnetic spectrum in terms of wavelength and photon energies. The visible region extends from 400 nm to 700 nm (wavelengths) with the corresponding energies of 3.1 to 1.8 electron volts (eV).

6.5 The Color of Objects

Consider a blue coat illuminated by white sunlight on a sunny day. The color of the coat is produced by the absorption of selected wavelengths of light by the dyes in the coat. Objects can be thought of as absorbing all colors except the colors of their appearance, which are reflected, as illustrated in Fig. 6.7. The blue coat illuminated by white light absorbs most of the wavelengths except those corresponding to blue light; the coat reflects these blue wavelengths.

If the blue coat is illuminated by red light (by pure red laser light, for example) the coat will appear black. If you try this experiment yourself on a *shiny* blue object, you will see the object (or the laser spot) as red because it will have a varnish that reflects some of the incident light. In the case of the coat, which probably would not have a varnish, you might still see red because the dyes in the coat cannot produce a perfect blue color (most blue objects have an admixture of red and other colors). These impurities in the coat's blue dye will reflect red light.

To emphasize the diversity of color, students in the Art Department at Arizona State University created a matrix of 88 shades of green. They were then asked, Which color is green? For further emphasis, they drew a still life of objects with eight shades of green. Color (and beauty) is truly and literally in the

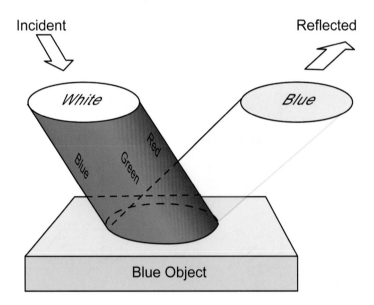

Fig. 6.7 White light composed of all wavelengths of visible light incident on a pure blue object. Only blue light is reflected from the surface.

eye of the beholder. In later chapters, you will find that color can be produced by mechanisms other than absorption and reflection (such as thin film interference), which create the colors of oil slicks and bubbles.

6.6 Sources of Light

As we've seen, the visible spectrum of light is just a small part of the entire electromagnetic spectrum, which extends from radio waves (long waves, kilometers in extent) to gamma rays (short waves, 10^{-3} meter down to 10^{-15} m — the size of the nucleus). X-rays have wavelengths of around 10^{-10} m, roughly the size of atoms. But where does this energy come from? How is light generated, and why isn't all light visible light? Let's start by looking at the Sun. (Figuratively, that is.) Ground-based and orbiting solar telescopes provide detailed views of the Sun and its myriad subtleties, including the cool sunspots seen in the visible light photo of Fig. 6.8.

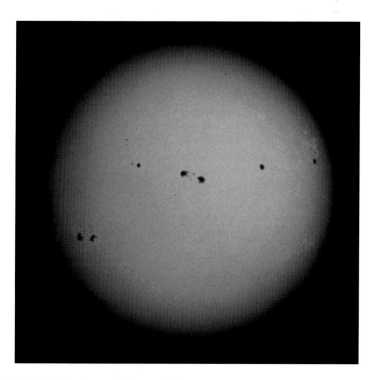

Fig. 6.8 Visible light picture of the sun with sunspots.

We look out from a tall building toward the horizon on a clear day and see thousands of colors, including various shades of brown soil, steel-gray buildings, green plants, azure sky, and a scarlet coat. Are these items giving off their own light?

No. All the light that reaches our eyes from outside during the day is reflected sunlight. Surely the light from the Sun must be white — there seems to be no preferential tinge to the objects we see. The blue sky? It's a result of light scattering by particles in the atmosphere, purely a terrestrial phenomenon. So where did all this sunlight come from?

Basically, it comes from inside the Sun. The complete answer involves complex nuclear physics, but a look at the basic principles will suffice for our discussion. Hydrogen atoms (of which the Sun is primarily made) collide in the interior of the Sun, where it is about a billion degrees C. At these temperatures, things move fast, and when atoms collide, they release tremendous energy, as scientists discovered (and governments learned) in the 1930s and '40s. Much of this collision energy in the Sun is given off as electromagnetic radiation — that is, light. But it's not visible light. In the interior of the Sun, the radiation generated is mostly *gamma rays*. Only through the incredibly long journey through the gaseous solar atmosphere are the gamma rays converted to lower forms of energy. By colliding with other atoms and giving some of its energy to them, the high-energy photon (or wave) eventually becomes an ultraviolet photon, then a visible light photon, and probably even an infrared photon, all of which are generated prodigiously in the solar atmosphere. Every beam of light, every photon from the Sun, was generated from the nearly countless numbers of photon-atom collisions that came before it. Each photon striking our eye as we look up has billions of ancestors, some existing only for a fraction of an instant, but all related to a photon created in a tiny fusion reaction, a nuclear explosion in the Sun's interior. Every second in that unimaginable furnace there are about 1 followed by 40 zeros of these photon births.

Here on Earth, light is created in much less energetic environments. All objects — a shoe, an electric stove, a flashlight, a little brother — give off light through the dance of electrons, the tiny particles inside atoms. If an object is hot enough and the electron dance is fast enough, a portion of that light will be in the visible spectrum. Often, as we'll see in Chapter 8, electrons are not bound to atoms and can roam freely or move about quite deliberately when exposed to a voltage. If you have enough free electrons, enough atoms, and enough voltage, you can reach a point where atoms and electrons are colliding so furiously that some of the released energy is given off as light. Let's think about what happens when we turn on a stove or a flashlight. An electric current is sent through the *filament* (usually a metal such as tungsten), and the free electrons perform a chaotic dance with the atoms, colliding, ejecting bound electrons, and recombining with atoms. Released energy in the form of light and heat is the end product of these processes. Both the stove and the flashlight (and other light bulbs) produce more infrared radiation than visible light radiation. To keep the tungsten filament from burning or melting, the glass bulb is filled with a mixture of argon and nitrogen gases that does not react with tungsten. When a *candle* burns, the combustion process combines carbon atoms (from the wick and wax) with oxygen to form carbon dioxide, heat, and light — again mostly in the infrared, but some of which is in the visible regime.

The blue sky and red sunsets provide a source of light associated with electrons bound to their atoms. A photon can *scatter* from an atom by colliding and raising the internal energy of an atomic electron. When this electron jumps back down in energy, it must release some energy to obey the principle of energy conservation, and it does so in the form of a photon. Sometimes that photon is of exactly the same energy as its progenitor (a clear glass window is one example of this process), or sometimes the photon is less energetic (a fluorescent light bulb is one example).

In contrast to light bulbs' *incandescence*, the notorious fluorescent bulbs of office ceilings operate as a gas discharge tube — a tube filled with an inert gas (usually mercury) and connected by electrodes to an alternating current (AC) source. The AC source shoots electrons through the gas, ripping the bound electrons from the gas atoms in a process called *ionization*. Other free electrons fill the void left by the ejected electrons and, like an atomic barter system, the atom emits a photon. Because the energy of the ionized electron is so high, the emitted photon, too, must be energetic (it is usually a UV photon). The inside of the tube is coated with *phosphors*, which absorb the UV photons and, through a process called *fluorescence*, produce visible light photons. At the same time, the electrons of the phosphor, as if they are scattering, jump down in energy levels from their high-energy state. (Recall that visible light is less energetic than ultraviolet light.) Like the incandescent bulb, the fluorescent bulb can produce many wavelengths of visible light, and so we see the light as white, which is all wavelengths mixed together. Consequently, fluorescent tubes have a broad energy spectrum similar to the tungsten lamp, and a series of sharp intense peaks in the red, blue, and green. The intense green line (mercury vapor) at 546 nm is sometimes used to calibrate *spectrometers*, devices that measure the spectrum of light. Neon lights in shop windows are gas discharge tubes (like fluorescent tubes) with the phosphor removed. Their color variance comes from the different types of inert gases used: neon emits red-orange light; argon emits blue light; and mercury, although not an inert gas, emits green light.

6.7 Replacing Edison

Fluorescent bulbs are barely a century old, and there are many light sources that are even younger. Light-emitting diodes (LEDs) and lasers were invented in the post-World War II technology boom and have had a great impact on our society

Fig. 6.9 Workmen install LEDs in traffic lights.

and culture. We will review LEDs in detail in Chapter 8, but, briefly, an LED emits light of a single, specific wavelength when electrons recombine with atoms in a semiconductor. Like other processes of light production (scattering, fluorescence, incandescence), electron recombination in LEDs involves the change in motion and energy of electrons. When electrons *recombine* (or rejoin) with an atom that briefly had an electron-vacancy, the electron gives up its energy by emitting another type of energy, a photon. This photon has a very specific energy (that energy which the electron gave up) and thus a very specific wavelength.

Red LEDs (long wavelengths) were first introduced in the late 1960s and have been used since then to light myriad electronic components, including stereos, appliances and toys. Shorter wavelength LEDs came later (blue LED was invented in the late 1980s), so that LEDs of all three primary-colored lights are now available. You may have seen the bright traffic lights (Fig. 6.9) installed in many cities; the new lights contain LEDs in place of outdated incandescent bulbs behind red, yellow, or green filters.

6.8 Revolution in White Light Sources

The normal incandescent light bulb is too hot to touch, and since most of the electrical energy consumed by the bulb is used to heat the filament, incandescent lights are not very energy efficient. Lighting from both incandescent and

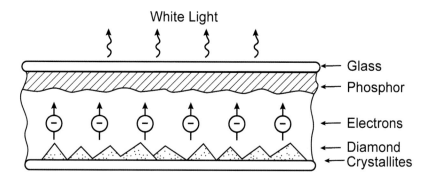

Fig. 6.10 Thin layers of diamond crystallites emit electrons that strike a phosphorescent screen. The phosphor absorbs the electrons and emits white light in the form of photons. (Courtesy Akio Hiraki, Kochi Technical University, Kochi, Japan)

fluorescent sources is responsible for roughly 20% of total electricity consumption. There are two developments underway that could have a major effect on lighting technology, reducing power consumption by 10% or more.

The first development comes from Kochi Technical University in Japan, where there is strong interest in producing thin layers of crystallites of diamond. The crystallites are very efficient emitters of electrons. Akio Hiraki, a lead researcher at the university, realized that he could make a very efficient light source by having electrons strike a light-producing phosphor, as shown schematically in Fig. 6.10. These phosphors are similar to those used in fluorescent lights. Hiraki also has shown that, by changing phosphors, he can make an energy-efficient light for growing hothouse plants without overheating them. The other development involves a team of twenty-five researchers at Sandia National Laboratories. They are producing semiconductor LEDs that emit light in the ultraviolet part of the spectrum. The UV light strikes a phosphor chosen to emit white light when exposed to UV. The light source is cool (as well as being energy efficient) and can be held in the hand. Try doing that with a 100-watt bulb!

It is still too early to tell which system will be the first to enter full-scale mass production, but no matter which it is, keep an eye out for big energy savings lighting the future.

6.9 Tricking Photons with Lasers

The laser has been around slightly longer than the LED, and it has seen just as many practical applications in its short lifetime: medical, military, information processing, and entertainment, to name a few. Like the LED, the laser creates light by harnessing and releasing the energy of electrons and photons. It does this, though, by tricking the electrons a bit.

In many materials, a given electron can occupy one of many energy states — the lowest energy is called the *ground state*. If an atom's energy is given a slight nudge, the energy of the electron can be raised, as we saw earlier with scattering. The material inside a laser (ruby, helium-neon, gallium arsenide) permits its electrons to be raised in energy (an *excited* state), then fall into a slightly lower energy state. In this intermediate state (say, 2 eV above the ground state), the electron may stay for a long time compared to the decay times of other energy states. If the electron were to spontaneously fall from this intermediate state to the ground state, it would release a photon with energy corresponding to the excited state (2 eV). Not a big deal. But if that photon strikes another atom whose electron is in the intermediate excited state (2 eV), the photon will stimulate the electron to go to the ground state and release another 2 eV photon. And so on. The photons that are stimulated are coherent waves, which means that they travel together in the same direction in lock step, like marching troops.

The first trick is to put as many electrons as possible in the intermediate state, and then let the material spontaneously generate some photons and, hence, stimulate emission of all the others, yielding a cascade of photons.[1] The next step is to sandwich the material between two mirrors so that the coherent light cascade will bounce back and forth, growing in strength as more and more 2 eV (for instance) photons are generated due to stimulated decay of those intermediate state electrons. If we allow 1% of the photons to escape from one mirror, we will have created light amplification by stimulated emission of radiation, a beam of photons all of one wavelength and polarization. (See Chapter 10 for a discussion of polarization.)

6.10 Structural Color

The final color-producing mechanism that we cover in this chapter is for the birds. And the butterflies. In 1869, a scientist named John Tyndall demonstrated that very fine dust particles preferentially scattered away blue light when illuminated with white light, much like gas particles in the atmosphere. This *Tyndall structural color* also is responsible for the color of a bluebird's feathers and for the way a dark-haired man's freshly shaven face will have a somewhat blue tint. The bluebird's color is not due to pigments in the feathers, but to tiny cavities of air in the feather strands that are larger than the wavelength of white light. These cavities act as scatterers, much like Tyndall's dust particles. Similarly, the fresh shave exposes keratin particles on the skin's surface, and these scatter the blue

---

[1] We put the material in the continuous intermediate state by using an external light source or an electric current.

from the white light of the bathroom lamp. In darker skin, the layers of skin underneath the first few layers absorb the other colors as well, leading to an even more pronounced effect.

Tyndall noticed and demonstrated structural color, but it was Lord Rayleigh who explained the phenomenon in terms of the preferential scattering of light by particles smaller than the wavelength of light. We will further explore Rayleigh's contribution to the understanding of nature's colors in Chapter 9 when we look at blue skies, white clouds, and grains of salt.

6.11 The Eye and Color Sensation

Our perception of color arises from the composition of light — the energy spectrum of photons — that enters the eye. The retina on the inner surface of the back of the eye (Fig. 6.11) contains *photosensitive* cells. These cells contain pigments that absorb visible light. There are two classes of photosensitive cells, called *rods* and *cones*. Cones allow us to distinguish between colors, and rods are effective in dim light and respond to differences in light intensity — the flux of incident photons — not photon energy. In dim light we perceive colored objects as shades of gray, not shades of color.

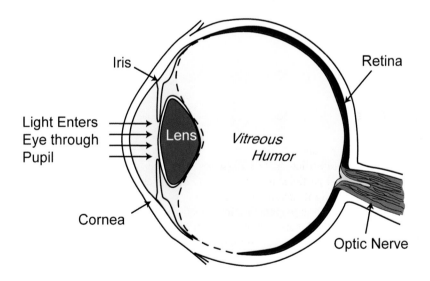

Fig. 6.11 A cross-sectional representation of the eye showing light entering through the pupil. The photosensitive cells (cones and rods) are located in the retina. Cones respond to different colors, and rods respond to differences in light intensity.

Color is perceived in the retina by three sets of cones, all of which are photo-receptors. They are sensitive even to those photons whose energy broadly over-laps the blue, green, and red portions of the spectrum. Color vision is possible because the sets of cones differ from each other in their sensitivity to photon energy. The maximum sensitivity is to yellow light, but the sensitivities of the three cones overlap. For every color signal or flux of photons reaching the eye, some ratio of response within the three types of cones is triggered. It is this ratio that permits the perception of a particular color.

In Chapter 7 we will examine two other color sources in nature's vast palette of light and color: diffraction and interference.

7 Diffraction and Interference

7.1 Light as a Wave

What if an object is not painted or dyed a certain color and doesn't generate its own light and spectrum? How can an oil slick, the wings of a butterfly, a thin sheet of plastic, or the corona around the moon be colorful? We'll see in Chapters 8 and 9 why rainbows, blue skies, and white clouds are colored, but those ephemera obviously have something to do with the reflection of sunlight. The two modes of color creation we'll examine here depend somewhat on sunlight and reflection, but as distinct phenomena we tie them to the more abstract notion of the wave nature of light. In this chapter we will truly break away from Newton's idea of light as particle and dive into the description of light as a wave.

7.2 Wave Interference

We played fast and loose with the terms *diffraction* and *interference* in earlier chapters. Here we give them formal definitions and a description of their effects. *Interference* is simply what happens when a wave interacts with itself or with another, similar wave. *Diffraction* is the bending of light that occurs when a beam of light travels around a corner or through a small opening. As we'll see, when the opening is small enough, the diffraction of waves produces interference effects that we can see and measure. When waves interact, they produce areas of constructive reinforcement and areas of destructive cancellation, regardless of the type of wave — water, sound or light.

Think of the large waves generated when swimmers splash and play in a pool. You may only notice when the wave crests come together to make a bigger wave, but a wave crest and a trough (the "dip" in the wave) can also meet and cancel out the wave momentarily; this destructive interference produces a temporarily smaller wave or even a flat spot. Constructive interference requires that the waves be *in phase*; that is, that their sources must generate waves in sync. Destructive interference occurs when waves are *out of phase*, or out of sync.

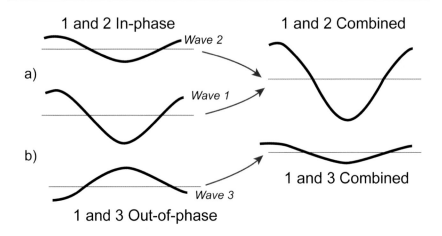

Fig. 7.1(a) Two sources oscillating in phase emit two waves that are everywhere in phase, and thus interfere constructively. **(b)** Two out-of-phase sources emit waves that interfere destructively everywhere. (From Falk et al. 1986)

Back to the pool analogy, imagine two swimmers each using a paddleboard to make waves. If they pump the paddleboards up and down at the same time, the waves will be in phase and produce large, constructively interfered waves when they meet, as in Fig. 7.1(a). If instead the wavemakers push the paddleboards exactly out of sync — one going up while the other is pushing down — the crest of one wave will meet the trough of another and they will cancel each other out, or at least make the resulting wave smaller, as in Fig. 7.1(b).

What are the consequences of these interactions for light waves? At the frequency (or wavelength, if you prefer) at which the light waves interfere, there will be patterns of alternating light and dark patches corresponding to where the waves constructively and destructively are interacting, respectively. Usually these types of patterns are seen when light is projected onto a screen or measured by a light-sensing detector, but you can see a similar pattern yourself with just your fingers. Use your thumb and index finger to create a narrow slit, as if you are trying to hold a tiny insect. Now bring the slit near your eye and look through it at a bright light source, such as a table lamp or the daytime sky. Bring your fingers as close together as possible without touching, and notice the tiny lines of light and dark running parallel to your fingertips. This is the interference of the light passing through the slit made by your fingers.

If light waves can interfere when they interact, why then are we not continually observing interference effects in midair from, for instance, two neighboring street lamps? As we saw in the previous chapter and will discuss in more detail in Chapter 11, white light is produced by the oscillations of electrons when atoms are heated and collide. As these electrons do not necessarily oscillate in phase with each other, the light emitted from the streetlamp is not *coherent*, or in phase. In fact, the phase of the streetlamp light changes constantly because the billions of electrons are moving in such random motions that successive light waves generated by those electrons change phase from one instant to the next. Any interference effects that might have been seen from two streetlamps would be immediately shifted due to the constantly changing phase of the waves. Only when the two (or more) sources are very close together or when they are coherent, such as in a split laser beam, can we observe interference effects.

7.3 Young's Interference

In the first few years of the nineteenth century, a London doctor named Thomas Young delivered a series of lectures to the Royal Society and to the public about discoveries he had made regarding the wave nature of light, a theory not exactly in vogue in the post-Newtonian haze of Enlightenment physics. The *emissionists*, Newton's followers, believed that light was best described as a stream of particles bouncing to and fro, attracted to or repulsed by matter. With a series of deceptively simple experiments, Young effectively negated all arguments against the wave nature of light and, with the important later contributions of Fresnel, relegated the emissionist theory to a historical footnote.

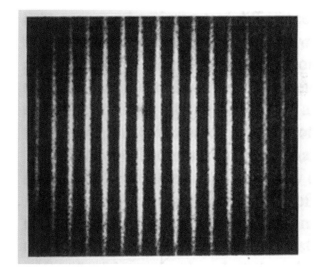

Fig. 7.2 Light intensity recorded on a screen behind two small holes illuminated by a single small light source. The dark stripes show where the light waves from the two sources interfere destructively, and the white stripes show where light waves interfere constructively. (From Sobel 1987)

In the simplest experiment, Young shined a light source at a screen etched with two narrow, closely spaced slits. On a screen some distance away, he viewed the results: a pattern of alternating dark and light bands, as shown in Fig. 7.2. Young said this pattern was the result of the interference of the light waves as they emerged from the identical slits, each acting as an identical light source. The dark bands showed where the waves destructively interfered and canceled; the light bands showed where they constructively interfered and added together. The emissionists concocted bizarre explanations to support their theory. The wave theorists pointed to the waves in a pond of water to support theirs.

7.4 Color from Interference

How, though, does interference of light waves produce colors, such as those in a beautiful peacock feather or in an oil slick on a wet road? We will review the mechanisms of thin film interference (oil slicks and soap bubbles) and iridescence (peacock and hummingbird feathers) below, but generally, interference produces colors when those particular frequencies constructively interfere and the other colors in the spectrum do not. So in an oil slick, blue is the predominant color where the thickness of the oil film has promoted the blue frequencies of the light to interfere constructively and all the others have canceled out. But right next to the blue spot is the green spot — the place where the film favors constructive interference of green frequencies — and all others, including blue,

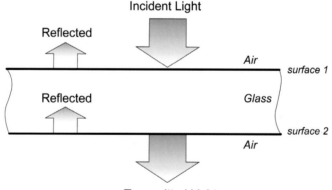

Fig. 7.3 A glass layer or film reflects some light (about 4%) at each air/glass surface. If the film is of the proper thickness, the light reflected from surface 2 will be in phase with the light reflected at surface 1, and constructive interference will occur for the reflected light at a specific wavelength.

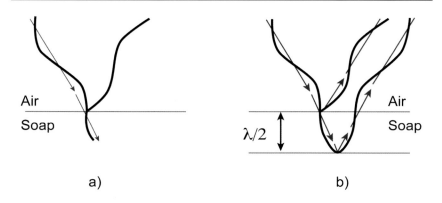

Fig. 7.4 Light incident on and reflecting from (**a**) soap and (**b**) a film of soap with half wavelength thickness. If the film is one-quarter wavelength thick, the reflected light will interfere constructively.

have destructively interfered. Interference colors are what remain after all other colors from a patch of film or feather have canceled out.

7.5 Soap Bubbles

Interference colors appear on soap bubbles, oil slicks, and oxidation layers on metals. These effects occur not because the light source producing the color generates coherent light (a colorful oil slick is seen in daylight, after all, and the sun is *not* a coherent light source) but because the light shines on two closely spaced surfaces: the front and back surface of a thin film of transparent material. Each of these surfaces acts as a new coherent light source, somewhat like the two slits Young used to demonstrate wave interference, and produces colorful interference effects, even though the soap or oil or oxides are themselves color-less. Figure 7.3 demonstrates how a light beam incident on a transparent thin film produces two reflected beams, one from the front and one from the back surface.[1] These two beams can interfere with each other to enhance or reduce the light intensity of a particular frequency (or wavelength) of the incoming light.

Like a wave on a rope tied to a wall, light reflected at the front surface of the thin film undergoes a phase reversal, Fig. 7.4(a), so that the troughs and crests switch. (If the crest of a wave on the rope points skyward and hits the wall, it will reverse and come back to you pointing toward the ground.) For a thin film in air, e.g., a soap bubble, the reflected light does not experience a phase rever-sal at the second (or back) surface, and the wave travels back through the film,

---

[1] You can substitute oil or soap or silicon oxide for the glass illustrated in the figure.

Fig. 7.4(b). When the film is one-half wavelength thick, as in the figure, the wave reflected from the back surface will emerge *out of phase* with the light reflected from the front surface. Destructive interference results, and reflected light at that wavelength is not seen. When the film is one-quarter wavelength thick, however, the waves emerging from the film will be *in phase* with each other and constructive interference results. Light at the wavelength (or color) corresponding to the appropriate thickness of the film will be reinforced and bright.

You can easily see the dependence on film thickness in a soap bubble. Notice how the colors swirl and dance as gravity and air currents affect the thickness of the soap film. At one moment, a patch of film will be the right thickness to reinforce the sunlight's red wavelengths, while in the next moment, the film thickness will change and green will be bright, then blue. When the film is too thin, all the colors interfere destructively and very little reflection is seen. On the other hand, when the film is too thick, all the colors interfere *constructively* and white light is reflected. This is the case for a pane of glass; it is too thick to produce interference colors, so it transmits and reflects white light.

7.6 Oil Slicks and Lens Coatings

If a thin film rests on another medium such as glass or water, the film thickness needed to produce interference colors will differ from that for thin films in air. The most mundane (but still striking) instances of thin film interference are in oil slicks on water and oxides on metal. Figure 7.5 shows an oil slick on a rainy street in Washington, D.C. The colors arrange themselves in rings because the oil film's thickness is consistent in circular patterns. You can perform this demonstration at home with lightweight oils (even chicken broth). It doesn't take very much oil to create a film; interference effects become apparent for films near 500 nm, or 0.5 microns, in thickness. Experiment with the amount of oil; if the film is too thick or thin, no interference will be seen, as noted in the previous section.

Boil pasta in an aluminum pot with salted water, then drain the water and view the intricate interference colors in the pot's interior. The colors emerge as the water evaporates and a thin layer of sodium oxide forms on the metal surface. You can also see this effect on a wafer of silicon as its surface layers oxidize and produce colorful displays of interference. Like soap, oil, and glass, these thin, metallic layers are semi-transparent to light and thus generate two coherent light sources at the front and back surfaces of the film.

As strange as it seems, adding a thin layer of transparent film of the right thickness enhances the transmission of light and reduces reflections on glass lenses. At just the right thickness for visible light, almost none of the light will

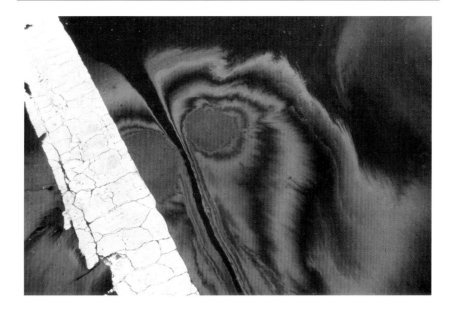

Fig. 7.5 An oil slick on a rainy street. (Photo by Steve Beeson)

be reflected, so nearly 100% will be transmitted through the glass. Engineers perfected these anti-reflection coatings in the early part of the twentieth century and thus advanced the use of multi-lens optical devices such as cameras and telescopes. Without the thin film coatings on camera lenses, internal reflections between the lenses would ruin the photographic image, and telescopes need as much light transmission as possible to work effectively. Look in the lens of your SLR (single-lens reflex, a misnomer) camera and you'll notice almost no reflection from its surface. The purple tint results from the anti-reflection coating being optimized to the middle of the visible light spectrum, in the green. The red and blue "wings" of the spectrum are not entirely canceled out and thus reflect back a small fraction of the red and blue wavelengths of the incident light. Combined, the reflected colors appear purple or violet.

The opposite effect, near 100% reflection, can be achieved by stacking multiple layers on top of each other (Fig. 7.6). The thin film layers have alternating high and low indices of refraction so that the reflections at each surface are in phase for one wavelength.

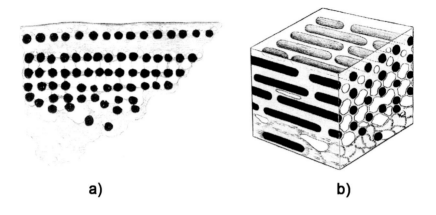

a)   b)

Fig. 7.8 An electron microscope view of the regularly spaced rods in a peacock feather. **(a)** Portions of cross sections through a peacock train feather barbule, (From an electron micrograph, magnification 40,000:1). **(b)** Diagram of a small portion of a cross section through a peacock barbule. Note the evenly spaced layers of melanin rods. Taken from H. Simon, *The Splendor of Iridescence.* Dodd, Mead and Company, New York, 1971

7.9 Diffraction

We return now to a topic introduced at the beginning of the chapter as the interference effects caused by light traveling through small openings or around sharp corners. Unlike interference due to layers of thin films, these diffraction patterns are caused by a wave interfering with itself as it spreads out through the small opening or slit. If we recall Young's double-slit experiment, we know how a light shined through two thin slits creates interference patterns on a distant screen as each slit acts as a new source of light waves, spreading outward and interfering with the others. Similarly, we can now imagine a single slit divided up into tiny areas, each of which we can consider a source of light spreading outward in phase. This is *Huygens' Principle*, named after the Dutch scientist introduced in Chapters 4 and 6.

When we add up the contributions due to interference of all the waves emitted by those tiny sources, we see on a distant screen[2] a pattern of alternating dark and bright spots. The spacing of these fringes is based on the wavelength of the light shining on the slit — if it is white light of all colors, different colors will appear to either side of the center, depending on their distance from the

---

[2] Distant compared to the size of the slit.

Fig. 7.9 A peacock feather. (Photo by Steve Beeson)

central bright spot. Like a prism, the slit produces a continuous spectrum of color, yet it is a spectrum caused not by the refraction of light through glass or plastic, but by the interference of the light waves through the slit and beyond.

7.10 Diffraction Gratings

We now see that interference slits and diffraction can be used to generate spectra of light sources. Many scientists in fact prefer to use diffraction gratings rather than prisms to produce light spectra because the gratings can be made much smaller than prisms, and they don't experience a prism's light loss. There are generally two types of diffraction gratings: transmission gratings and reflection gratings. As the names suggest, each produces a light spectrum based on the way light interacts with the grating surface. Transmission gratings consist of tiny slits etched into an opaque surface in a regular pattern. As in Young's slits, the light wave passes through the slits and interferes with itself and other new wave fronts beyond the grating. Short of buying a diffraction grating from a science supply store, you can find rough transmission gratings in some types of window screens and in the frost on a window. In either case, observe a distant bright light source (such as a street lamp) through the window or screen and look for a ring of light colored like a rainbow around the lamp. As long as the frost is thin

Fig. 7.10 Reflection grating colors shown on a compact disc. (Photo by Don Thompson)

and evenly distributed, you should see a nice ring around the lamp. Similarly, the window screen diffraction requires a fine, tight mesh of wires, the holes of which provide the slits.

In a reflection grating, light shines on a plate etched with regularly spaced grooves, as in a phonograph.[3] The grooves act to break up the wave fronts, like the opaque material between slits, and the reflected light diffracts, producing a spectrum in the direction from which the light came. You can see a diffraction pattern in an LP or single phonograph with a distant bright light source shining on the disc. Or better yet, you can use a compact disc to see diffraction. Observe the rainbow of colors when you view sunlight reflected from the back of a CD (the side opposite the label); the pits in the polycarbonate plastic from which the disc is made are spaced about 1.5 microns apart. When white light is incident, the regularly spaced pits diffract and disperse the reflected light waves, as in Fig. 7.10. If you have a pen laser, you can observe the diffraction of monochromatic (one-color) light from the CD surface: Shine the laser on the CD and aim the reflected light at a white wall or paper. The distance of the reflected spots from each other tells you how closely the slits, or pits, are spaced.

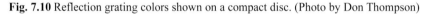

[3] CDs (compact discs) act as reflection gratings. This is especially true for newer discs, which have a higher information capacity and more grooves per inch.

8 Rainbows

8.1 Through the Looking Glass

White light is composed of all wavelengths of visible light. The waves of
sunlight streaming into a room or coming from a table lamp are a mixture of all
the colors of the spectrum. Optical devices called prisms split white light into its
various wavelengths, or colors, each of which responds to the glass in a different way. The degree to which light bends, or refracts, at a surface depends on
the wavelength, or frequency, of the light: blue light tends to refract more than
green light, and green light refracts more than red. This effect, called *dispersion,*
occurs in all substances that bend light.

Try this experiment at home: fill a cylindrical drinking glass with water (or,
better yet, a spherical fish bowl or vase) and place it on a table in a ray of sun-
shine. On the side of the water-filled vessel opposite the sun, you should see a
rainbow spread out on the table — maybe not as majestic as a rainbow in the sky,
but a spectrum still. The glass of water, like a prism, refracts and disperses light.

8.2 The Pot of Gold

Think back for a moment to the last time you saw a rainbow in the sky. Where
were you standing relative to the sun? Were you looking toward the sun, or
away from it? As you stood looking at the rainbow, the sun must have been
above and behind you. The rain was somewhere in front of you, if not falling on
you at the time. Since the sun was behind and the rain in front of you, the
sunlight must have been bouncing off the raindrops and reflecting into your
eyes. This concept contradicts the prevailing theories of rainbows throughout
most of history — Aristotle believed the rainbow was reflected from a distant
cloud. Other cultures believe rainbows to be tangible entities, colorful snakes or
bridges upon which gods ascended or descended from their heavenly thrones
(Lee et al. 2001). It wasn't until people discovered that there was no "end of the

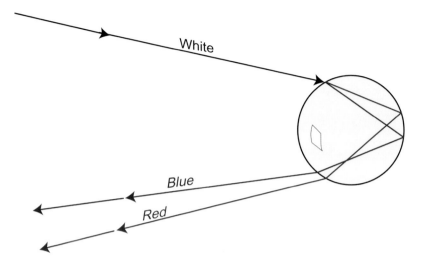

Fig. 8.1 Raindrops disperse white light into its component colors and reflect the light towards the ground. A ray of blue and a ray of red are shown here. The rest of the spectrum would lie between these two light rays.

rainbow" and, hence, no deity or gold there, that its description as a purely optical phenomenon took root with the Greeks.

Even though there is no gold at the end of (or even in) the rainbow, it still typically has lots of colors. If the sunlight is primarily white light, why do we see colors in a rainbow? The colors emerge because the sunlight is not only reflecting from the raindrops, it is also refracting and dispersing in the raindrops, as you saw with your glass of water and as illustrated in Fig. 8.1.

Notice in the figure how the raindrop acts like a prism, separating the white light from the sun (shining in the upper left) into its component colors. For simplification, we have only shown the red and blue beams, which are the extremes of the spectrum; the other colors in the spectrum are emitted in sequence between red and blue. Notice also how the raindrop acts like a mirror in that it reflects (most of) the sunlight back toward the sun or down to the ground. These reflected and refracted rays are the ones that you see as a rainbow.

White light from the sun hits the raindrop and is dispersed as it enters, much as light is spilled as it passes through a prism. The separated colors are then reflected from the backside of the raindrop and exit the sunlit side, where they refract once again, due to the change of index of refraction between the water and air. Figure 8.2 shows the result of dispersion — each color emerges at a different angle: red light emerges at 42° and blue light emerges at 40.6° relative to the incoming ray of sunlight.

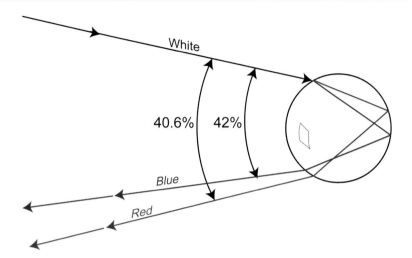

Fig. 8.2 A raindrop reflects dispersed light in a very small range of angles. Red light is reflected at 42° from the incoming sunlight; blue is reflected at about 41°. This explains why rainbows are so narrow on the sky: their brilliance and sublimity make them seem larger.

8.3 A Rainbow by Hand

Even though each raindrop creates a tiny spectrum of color, at any given moment it is contributing only one color to the rainbow that you see. How can this be, if rainbows have so many amazing colors?

We should ask first, what is the pattern of light displayed by a raindrop illuminated by sunlight? You can do a simple experiment if you have a round flask or vase made of clear glass, a flashlight, and a piece of white paper or posterboard. Cut a hole in the paper about as large as the vase's spherical bowl, and shine the light through the hole onto the water-filled vase. Look for a circular rainbow on the flask-side of the paper when you shine the light on the water-filled flask or vase. Why does the rainbow form a complete circle around the flask? The light isn't only shining on one little area of the raindrop; the flashlight illuminates the whole front side. White light shines on the left side, and colored light emerges on the right side. Similarly, white light shines on the right side and colored light emerges on the left side. This is happening at *every point* on the sunny side of the raindrop. The result is a circle of light created by the incoming parallel light beams reflecting and refracting through a spherical surface.

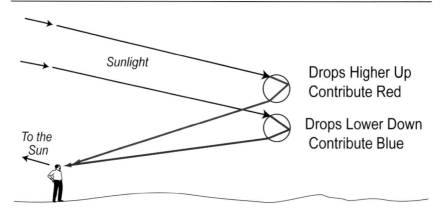

Fig. 8.3 Many raindrops contribute to a rainbow. The rainbow's red comes from raindrops higher in the sky than those that contribute to the green or blue.

Each raindrop is emitting the spectrum of colors in a circular, rainbow-like pattern, but surely a rainbow visible for miles is far too large to be created by just one raindrop. We find that we are seeing only one color (indeed, only one tiny beam of light) from each raindrop. While looking at a rainbow, we must be seeing the refracted light from thousands of raindrops. As it turns out, raindrops lower in the sky contribute blue and green light, and raindrops higher in the sky contribute red and yellow light, as shown in Fig. 8.3. How, then, is the bow formed from so many raindrops, each contributing a sliver of color? We find that geometry plays an important role in the creation of a rainbow.

8.4 The Antisolar Point

When you look at the ground on sunny day, the shadow of your head marks a spot called the *antisolar point*, the point 180° away from the sun, illustrated in Fig. 8.4. If the sun is in the sky, the antisolar point is *below* the horizon. If the sun has set, the antisolar point is *above* the horizon.

What does this tell us about rainbows? The antisolar point tells us where we can expect a rainbow to form, since the colored light from the raindrops exit those raindrops at specific angles that we can measure with respect to the antisolar point.[1] In other words, if we see a rainbow in the sky, it's because of raindrops that are 40.6° to 42° from the antisolar point, reflecting colored light into our eyes.

---

[1] We can measure from the antisolar point because the line pointing towards it is exactly parallel to the ray from the sun, as in Fig. 8.3.

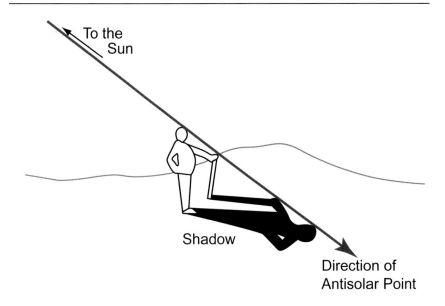

Fig. 8.4 The top of a shadow marks the direction of the antisolar point, the point, measured from your head, 180° from the Sun.

Let's step back and look at the larger picture. We have a block of raindrops in the sky. They are illuminated by the sun, which is refracting and dispersing the light into its various colors and reflecting those colors back. The red light reflects back at 42° relative to the incoming sunlight, which is our antisolar line. The blue light reflects at about 40° from our antisolar line. That is, *every* raindrop 42° from the antisolar line reflects red light back to our eyes. If we could look at *every* raindrop 42° from this line, we would see a *circle* of red-emitting raindrops centered on the line.

Similarly, every raindrop that is 40.6° from the antisolar line emits blue light into our eyes. Raindrops emitting orange, yellow, and green light are somewhere between 42° and 40.6° from the antisolar line. In an ideal world, we would see complete and perfect circular bands of the spectral sequence, from violet to red, called *rain circles* or *rain bands*. Unfortunately, the horizon gets in the way of most of the rain, so we only see an arc or a *bow*.

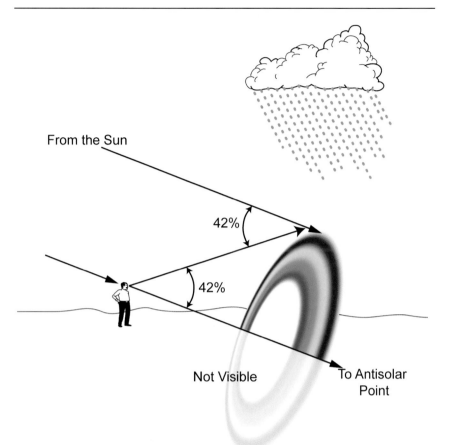

Fig. 8.5 The rainbow is an Earth-interrupted circle of every spectral color. Every rain-drop 42° from your antisolar point contributes red light. Every raindrop 40.6° from the antisolar point contributes blue light.

8.5 Rainbows in 3D

Rain does not fall in a flat sheet but in varying distances from you. This causes a rain circle to be formed at varying distances. The farther the raindrops are from your eye, the larger is the rain circle, because the radius of the circle subtending, say, 42°, gets larger as it gets farther from the reference point (your eye). Circles that occur at increasing distances and with uniformly increasing radii centered on a line create the shape of a cone. We find that the raindrops that contribute to your rainbow all lay on a *cone* with its apex at your eye. As clear and distinct as a rainbow may seem, it does not occupy a specific, two-dimensional region of space; when you are looking at a rainbow, you are looking at the collective light from many raindrops, some closer, and some farther from you, all of which, for a fleeting moment, produce a cone of color with its apex at your eye. If you move to the left or the right, you are looking at new raindrops, and hence, a new rainbow! If you are admiring it with a friend, you are both seeing different rainbows. A rainbow is your own.

8.6 The Double Rainbow

Every once in a while, if you're lucky, you may see a second rainbow on the outside of the first, brighter rainbow, as in Figs. 8.6 and 8.7. This is the *secondary rainbow*, which occurs when raindrops high in the atmosphere refract and reflect light back to the viewer. These raindrops are higher than those that cause the *primary rainbow* and are special because they internally reflect the incoming sunlight *twice*, unlike the primary raindrops that reflect sunlight just once.

Fig 8.6 A double rainbow. (Photo by Steve Beeson)

Fig 8.7 A double rainbow in Yosemite National Park. (Photo by Steve Beeson)

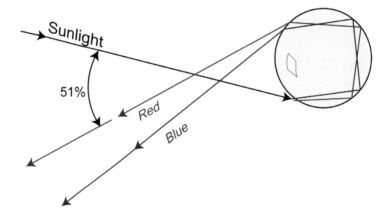

Fig. 8.8 Some raindrops doubly reflect sunlight in their interiors. If the light emerges at or near 51° from the incoming sunlight, it may contribute to the secondary rainbow.

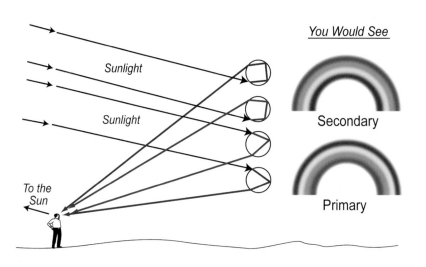

Fig. 8.9 The order of the red and blue contributions is switched in the secondary raindrops relative to the primary. Secondary rainbows are the inverse of primaries.

Fig 8.10 A rainbow at twilight. (Photo by Steve Beeson)

Notice in Fig. 8.8 that the sunlight hits the bottom of the raindrop to make the secondary rainbow. There is still sunlight hitting the other parts of the raindrop. However, some of the light is simply transmitted through the raindrop without reflecting, and some light is reflected and dispersed, but not into our eyes — that light is for someone else's rainbow.

Notice also in Fig. 8.6 that the sequence of colors in the secondary rainbow is switched compared to the primary bow. The explanation lies in comparing Fig. 8.1 to Fig. 8.8: the spectral sequence from a secondary raindrop is reversed relative to a primary raindrop. Another interesting phenomenon is visible in both Figs. 8.6 and 8.7: the light inside the rainbows is brighter than the light outside.

8.7 The Light inside a Rainbow

Sometimes when you look at a rainbow, the sky inside the bow looks brighter than the sky outside the bow. Why does this happen? Remember that a rainstorm is a three-dimensional object. Rain is falling from one or more large clouds onto the ground. The rainbow is a small area on the sky, and the sun shines on many other raindrops than those that make the rainbow.

Every raindrop reflects some of the sunlight without significantly refracting it. Some of the light is reflected at the front surface of the raindrop, some is reflected from the back if the angle of incidence is too small (smaller than the 42° associated with the rainbow). In fact, all the raindrops are reflecting some light at small angles, because light is incident on the entire front surface of the raindrop. However, only those raindrops within 42° of our antisolar line reflect light

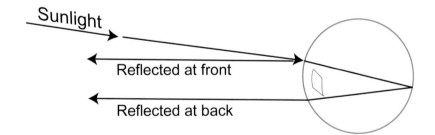

Fig. 8.11 Raindrops reflect sunlight at their front and rear surfaces. If the sunlight, the raindrop, and the observer are not in perfect alignment, you may just see reflected white sunlight, or no reflected light at all.

back to our eyes. We can think of the 42° of the rainbow as the upper limit at which the reflected light can hit our eyes. Any raindrops outside the cone will necessarily reflect light back, but the light will either go over our heads or past us on the right or left. Some of the raindrops *inside* our cone will reflect colors to our eyes, but most will simply reflect white light straight back to us or contribute to someone else's rainbow.

8.8 Rainbows Far Afield

Rainbows can also form in mist or fog illuminated by the sun. Look back to Fig. 8.7. A double rainbow, with white light inside the primary, forms in spray from Nevada Falls in Yosemite National Park. Waterfalls, lawn sprinklers, and mountaintop fog make excellent rainbows.

The rainbow in Fig. 8.10 appears in a twilight sky peppered with gray and white clouds. The clear sky between the clouds is a deep, dark blue.

In the next chapter we will look skyward again to examine the glorious spectrum of colors in the atmosphere: blue skies, white clouds, and red sunsets.

9 Sea, Sky, and Cloud

9.1 Beam of Light

Several times in this book we have discussed blue skies. Together with the sea, the sky is often our prime example, our default value, of blue. We know white clouds in a blue sky are beautiful. But why blue? Why white? To understand the reasons for the colors above and below us, we must first follow a beam of light as it strikes a partially transparent material, as in Fig. 9.1.

Every time light encounters a boundary between two different transparent materials, some of the light is reflected (about 4% reflection at the air/glass boundary), and some is transmitted with a change in direction due to refraction. If the first surface is smooth, a mirror-like or *specular* reflection will occur. If the surface is rough, *diffuse* reflection will occur. As it penetrates the material, light can be scattered or absorbed.

9.2 The Color of Sky

The light in the sky simply is scattered sunlight. By scattered, we mean it is re-flected by small particles, and in the clear sky, the small particles are the air molecules N_2, O_2, and others that are smaller than the wavelength of sunlight (400–700 nm). The scattering by these tiny particles is greater at the violet and blue end of the spectrum. Figure 9.2 shows the fate of a typical beam of sunlight in the atmosphere. Most of the blue light is scattered in all directions, some green light is scattered, and red is preferentially not scattered unless the light travels a great distance through the atmosphere (see Section 9.7).

The Sun emits more blue than violet, so we don't see violet skies, although violet is scattered the most. The reason blue is so pervasive in the sky is that the scattering has no preferential direction — like a rack of billiard balls struck by a well-aimed cue, the blue light scatters in all directions. Everywhere we look in a clear sky, we will see scattered blue light. The other colors similarly scatter, but in much lower quantities. The color of the sky, then, is barely violet, lots of blue

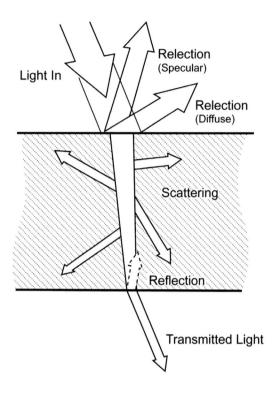

Fig. 9.1 The adventures of a beam of light passing through a block of partly transparent material. The light is reflected, scattered and transmitted.

and decreasing amounts of green, yellow, and red. The sum appears to our eyes as "sky blue." If you observe the sky with an optical spectrometer, you will note that the blue sky has a broad band of colors, extending from blue and fading toward red.

9.3 The Color of Sea

Many sky watchers mistakenly believe the sky's color is due to a reflection of the blue tint of the sea. This is not true, although the inverse is partly accurate. The color of the sea is attributed to three effects: reflection, scattering, and absorption. From the shore, looking far out over the ocean's surface, you see the sea's color as a reflection of the sky. Like the glass window in Chapter 2, at such great distances and angles, the sea's surface nearly perfectly reflects the blue

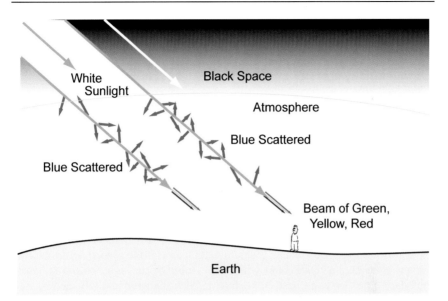

Fig. 9.2 Beams of white sunlight travel through space and encounter the Earth's atmosphere. The blue wavelengths are scattered in all directions by small molecules of air; the green, yellow, and red wavelengths are scattered to a lesser degree. As we look at the daytime sky, we see the scattered blue wavelengths of sunlight.

Fig. 9.3 A dolphin swims beside a dive boat in the Pacific. (Photo by Elizabeth Mayer)

sky, gray clouds, or red sunset, even in choppy water. Front surfaces of waves approaching you, however, may reflect or scatter the light coming from the depths of the ocean, which is a much darker color. If you look straight down into ocean, a different picture emerges, as seen in Fig. 9.3, where a dolphin splashes in the blue Pacific near Hawaii.

Like the atmosphere, the molecules of pure water scatter blue and violet light in all directions. Red, yellow, and orange wavelengths are preferentially spared from scattering, but, unlike their fate in air, they are usually *absorbed* by the water molecules. Greens are transmitted relatively well.

In sea colors, though, we must also take into account the seabed and its reflective properties. Water absorbs all the red wavelengths of a beam of white sunlight at a depth of about 30 m (~100 ft). If the ocean is deeper than 15 m, any light emitted from the ocean depths will be mostly bluish-green due primarily to the scattering, absorption, and reflection of the various wavelengths of white sunlight: blue is scattered, red is absorbed, and green is transmitted and sometimes reflected up. If the water is shallow, the color of the seabed makes an important contribution. Dark sand will absorb most of the light, so the sea may look dark blue or green (these colors having been scattered and reflected by water molecules). An ocean floor of white sand will reflect most of the light incident on it, and so, because of the scattered blue/green and reflected white, shallow water will appear aquamarine or light blue when viewed from above. The color transition between a shallow sea and a deeper one, or a sandy bed and a coral reef, is clearly visible in aerial images of tropical beaches and reefs, where the reddish flora creates a wine-colored sea.

9.4 The Color of Smoke

The same phenomenon occurs when other fine particles are illuminated by white light. In Fig. 9.4, we ask if smoke emitted from the end of a cigarette and smoke exhaled from the mouth are different colors. Observe the smoke rising from the lighted end of a cigar or cigarette — in ambient light, the smoke may take on a bluish tinge because the tiny particles of smoke preferentially scatter blue light. If you view a light source through the smoke, however, the smoke may appear yellow — a result of the blue wavelengths scattering out of the beam, leaving the green, red, and yellow light.

The exhaust from a diesel-burning engine often emerges blue before it dissipates. In humans, the blue eyes of an infant are due to the fibrous network of tissue in the iris scattering the blue in white light. As the infant develops pigment, the eye color may change to green or brown. (Those with little pigment development retain their blue eyes throughout adulthood.)

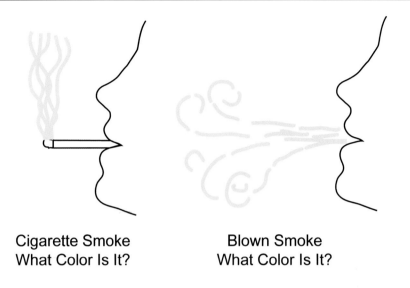

Cigarette Smoke
What Color Is It?

Blown Smoke
What Color Is It?

Fig. 9.4 In ambient light, tobacco smoke from cigars and cigarettes can change color after being inside a smoker's mouth.

9.5 White Clouds and Smoke

After observing the wispy blue smoke from a burning cigar, watch the smoke that is exhaled. Its white color is due to scattering again, but in this case, as with clouds, mist, and paper, the scattering particles are large compared to the wavelength of light. Why? Inside the humid mouth and lungs, moisture condenses on the smoke particles, making them much larger. Literally, a cloud of smoky water droplets is exhaled, as in Fig. 9.5.

But why is the cloud white? Like exhaled smoke, a cloud is simply the condensation of water on small grains of dust or smoke in the atmosphere. As the water condenses out of the air and onto the grain, the particles grow larger. These particles eventually become much larger than the atom-sized air molecules (N_2, O_2) that scatter the blue from sunlight. When white light falls on a group of these water droplets, the light scatters, reflects, and refracts through the particles, as in Fig. 9.6. Eventually it escapes, but no color has been preferentially scattered out. White light emerges at random from the cloud like static from an old TV on the wrong station. White clouds are the white noise of light.

Of course, not all clouds are white. Figure 9.7 shows clouds of many shades of gray and white. The gray clouds are in the shadow of other clouds and so do

Fig. 9.5 A Zimbabwean woman exhales a cloud of white smoke. Water from the mouth and lungs condensed on the smoke particles scatters light without preference for wavelength — all colors scatter equally, and white light is the result. This properly-bespectacled observer stands in the penumbra of the shadow of the Moon, which is blocking the Sun. (Photo courtesy of Reuters/Howard Burditt/Hulton/Archive)

not scatter white light. Some appear almost black when viewed with the white cloud in the background. Some clouds present a white or even a yellow haze-colored face to the sun while their backsides are in shadow. The cirrus clouds high in the sky appear white across the photograph — there are no other clouds to shade them from the sun's white light. The appearance of a *contrail* (condensation trail) in the sky signifies the passing of a jet aircraft whose particulate exhaust provided a seed for water in the upper atmosphere to quickly condense and create a cloud. Contrails often are the primary source of clouds in some skies.

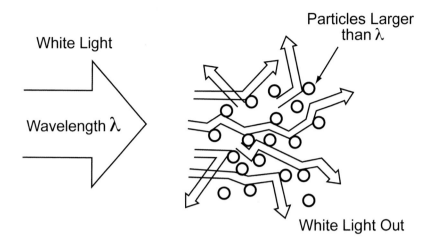

Fig. 9.6 Particles larger than the wavelength of light scatter all wavelengths equally. White light impinges on the particles, and white light is emitted.

Fig. 9.7 Contrail in the Arizona sky. (Photo by Steve Beeson)

9.6 Salt with your Beer?

You sit down at the *cabana* with a pint of cold beer and salted chips. Listen to the ocean pounding the land and watch the surf emerging from its watery journey as a glimmering white. All around you, sunlight is scattering from multiple surfaces and large particles to produce the white color of foam. In this sense, a pile of salt *is* foam; each salt crystal is transparent, like drops of water, but when put in a pile, the light scatters and reflects from many surfaces, emerging randomly as white light. The foam on the beach and the foam in the beer exhibit the same properties (through small bubbles of clear liquid), as do white fur and feathers, which have large air pockets that do the scattering.

Even colored glass and rock candy, when pulverized, look white. Figure 9.8 shows a piece of rock candy that's been smashed with a hammer. Notice the large, transparent-but-colored pieces and the white pile of granulated sugar. White light passes easily through the glass-like solid candy, but it gets caught up — multiply reflected — in the granules of crystalline sugar. The mixture and confusion of all the scattered wavelengths results in the perception of a white pile, though each particle is itself transparent.

Fig. 9.8 A transparent piece of rock candy becomes a pile of white sugar when pulverized. The small (but larger than the wavelength of light) pieces in the pile scatter white light in all directions. (Photo by Steve Beeson)

9.7 The Remains of the Day

At the end of your day at the beach, the sky is a glowing ember of red and orange in the west, and your thoughts are white foam from the beer. How does the sky turn from uniformly brilliant blue to fiery, bold red in a matter of minutes? Let's return to our beam of sunlight in Fig. 9.2 and its fate through a typical sky.

We recall that, as the beam travels through particles of air that are smaller than the wavelengths of sunlight, the violet, blue, and green photons are preferentially scattered out of the beam in all directions. What remains of the beam is mostly yellow, orange, and red, but this depends on how far the sunlight has traveled through the atmosphere. If coming from straight overhead, a shaft of

Fig. 9.9 Pink clouds reflect the end of the day.

sunlight retains much of its color and strikes the observer mostly as white light. If it travels, however, through a greater distance in the atmosphere, as at sunset or sunrise, it will scatter out more blue and green. In the early or late part of the day, the sunlight traverses many kilometers of atmosphere, and all that's left of the beam is yellow, orange and red. These warm colors illuminate foreground clouds as well, turning them pink, then orange, then blood red as the sun sinks farther below the horizon. In Fig. 9.9, clouds tinted pink reflect the end of daylight.

On a hazy or dusty day, the light of the sunset is scattered even more due to the presence of dust or water droplets in the atmosphere. In these cases, the twilight sky is often brilliantly red and orange in the direction of the setting sun — all of the violet, blue, green, and yellow is scattered out of the beam. Figure 9.10 shows a fiery sunset over Phoenix, enhanced by the hazy conditions of the city. If a volcanic eruption has ejected fine particles of ash high into the atmosphere, the sunsets for months or years afterwards can be especially dramatic, as was witnessed after the Mt. St. Helens, Agung, El Fuego, and countless other volcanic eruptions throughout recorded history (Meinel & Meinel 1983). If you hear of a large volcanic eruption along the Pacific Rim, look for spectacularly colored sunsets soon afterwards.

Sometimes a purple light is seen overhead or high in the western sky after sunset. This beautiful phenomenon is caused by a combination of reddened

Fig. 9.10 Haze enhances a Phoenix sunset.

sunlight scattered by air near the ground and blue light scattered from air high overhead that is still illuminated by the sun. As we recall from Chapter 5, the eye registers both red and blue light separately, and the brain interprets the combination as purple. Usually a very clear, cloudless atmosphere is needed to see this phenomenon. If the purple light is seen on snow, the effect is called *alpenglow*.

9.8 The Shadow in the East

If you turn away from the dying light in the west and look toward the east, you might see the shadow of a planet. Just as the sun is disappearing, turn around to look for an arc of purple light in the east (the *anti-twilight arch*), then a darker band below it, near the horizon. As the sun dips farther below the horizon, the dark band will rise higher; this is the Earth's shadow projected on the atmosphere. It is visible because some of the remaining sunlight is striking and scattering from the upper atmosphere, making the sky that is high in the east bright in contrast with the shadow. A few minutes later, though, your part of the Earth revolves beyond the point where scattered light, even from the upper atmosphere, will reach you; only a dim glow remains of the fire in the west, and Earth's shadow engulfs the dome of the sky. Without direct sunlight as the source of scattered light, the nighttime atmosphere becomes transparent, and we look out onto the void of space as it is sprinkled with scintillating points of starlight.

9.9 Beyond the Horizon

Examples of refraction can be found not only in pools of water, panes of glass or precious gems; the atmosphere provides plenty of chances to encounter refraction in a natural setting. Perhaps the simplest example occurs every day when the sun sets. Sunlight enters the atmosphere at a highly oblique angle at sunset. Because of this, as the sun seems to be approaching the horizon, we are actually looking at a sun that has already set. You can confirm this by noting that the sun appears to slow down as it sets; this could only happen if its light was being bent in the atmosphere. The refraction occurs smoothly at the tenuous boundary between atmosphere and empty space. Figure 9.11 shows how atmospheric refraction alters the apparent position of the sun by a full solar diameter.

At least one night a month you can view the setting sun and turn around to watch the rising moon in the East. Because of the refraction of the atmosphere, the full moon sometimes takes on an oblong shape near the horizon, flatter in the vertical direction but unaffected in the horizontal. The change in density, and hence, refractive index, of the layers of air near the horizon cause the light

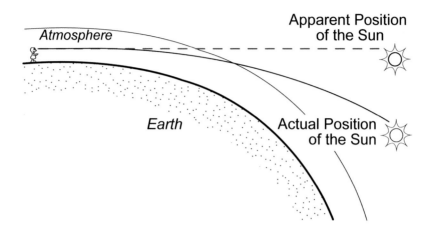

Fig. 9.11 Sunlight is bent as it enters the Earth's atmosphere from space, especially at sunset, when the incident angle is much greater. The sun can appear one full diameter above where it actually is.

passing close to the horizon to bend more than the light passing higher up (from the top of the rising or setting moon). A similar result can be seen with the setting sun, if intervening haze or clouds obscure some of the brightness of the solar disk.

Place yourself on a coast looking over the ocean to get the best view of a distorted sun. The ocean shore is also the ideal place to avoid succumbing to "moon enlargement syndrome." This phenomenon affects casual observers of a rising full moon who claim that it seems larger at the horizon than when high overhead. Many people like to invoke atmospheric refraction to explain this puzzlement. The cure is actually less complicated. Figure 9.12 compares the size of the lunar disk near the horizon, lurking among buildings and foliage, and overhead, away from terrestrial distractions. Each image was obtained with the camera lens set at the same focal length. We see that the Moon in fact does not get smaller as it rises in the sky. The enlargement of the Moon is a psychological effect caused by the superposition of everyday objects in our line of sight to it. How often do we see a rather large globe of light near a familiar building? Not very. So our mind tends to overcompensate and increase its perceived size.

Try this trick (if you're not easily embarrassed): stand with your back to the rising Moon, bend over at the waist, and view fair Luna from between your legs. From this perspective neither the upside-down buildings nor our celestial satellite seem as familiar, and the Moon is brought down to its rightful bulk. Observing the Moon over the ocean also tends to diminish the perception of increased size, as we view no earth-bound objects in comparison.

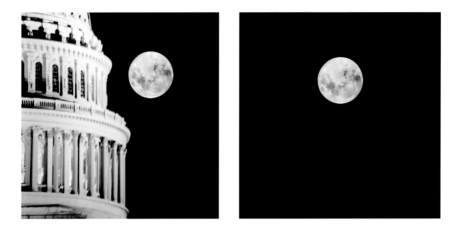

Fig. 9.12(a) The rising Moon **(b)** The Moon a few hours later on the same night. Both images were taken at the same lens focal length.

9.10 An Oasis in the Sahara (or the Arctic)

Gradual refraction is also responsible for *mirages*. These elusive, ephemeral images are caused by changes in the refractive indexes of adjacent layers of air. At 100° C, air has a refractive index of 1.0002, while at 0° C, it has a refractive index of 1.0003. Even this slight change in the optical density of air can cause light to bend toward cool air over long distances, as shown in Fig. 9.13. When a layer of cool air is above a layer of hot air, light originating from an object (e.g., the sky, a mountain) in the cool layer bends up away from the surface and creates an upside-down image of the object. Effectively, a slow total internal reflection occurs as the light passes through the air layers of different refractive indices.

On a long, lonely road through the desert, water seems to cover the road a mile ahead. As you approach it, though, the water disappears, and you just never seem to get to that cool oasis in the middle of the road. The "water" is a mirage of the sky (often blue) caused by the hot air above the road bending the light that passes through the cool air/hot air boundary. You're not used to seeing the sky on the road, so your mind associates this reflected light with water. It disappears because the layers directly in front of you can't bend the light so drastically — they need a long path length since the difference in the refractive indexes is so tiny. You also can see these types of mirages over warm water on a cool day: ships on the water or a shore in the distance may be distorted, rising far above the water surface or disappearing behind a wall of mirage'd sky.

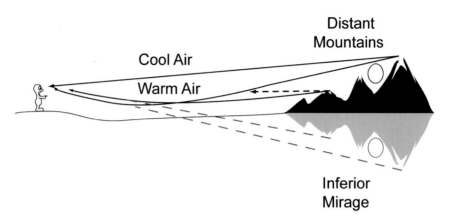

Fig. 9.13 An inferior mirage forms when cool air lies above warmer air. The change in optical density refracts light upward so that distant objects appear inverted.

If the warm and cool layers are reversed, you may see even stranger effects. In this case the light is refracted *downward* as it passes from cool air with a high refractive index to hot air with a low refractive index, as in Fig. 9.14. Because of this temperature-inversion effect, you can sometimes see objects that are below the horizon, as the light from them that would normally pass over your head is bent down to you.

This effect is thought to have caused numerous early explorers to attempt to reach lands that seemed only a few miles distant but were actually hundreds or thousands of miles away. Greenland and Iceland were possibly discovered with the help of these inverted, or *superior*, mirages that pointed Celtic and Scandinavian explorers in the direction of the faraway lands. Sometimes a mirage costs money. Commander Robert Peary, the famed Arctic explorer, reopened the 90-year-old mystery of *Crocker Land*, a distant, fantastic mountain range that was sometimes seen but never found. Peary claimed he re-spied the famous phantom range from a peak in northernmost Canada in 1909, and four years later, Commander Donald MacMillan set out to investigate the unexplored Crocker Land near the North Pole. Upon viewing the same landscape described by Peary, his exploration party pressed forward. What they found was polar ice, polar bears, and desolation, but no Crocker Land. As they left the Arctic ice, the explorers turned to see behind them the same elusive mountains and valleys that Peary had seen. But leaving was not as easy as arriving — MacMillan and his crew were stuck on the ice for two years, costing the American Museum of Natural History $100,000 to search for and eventually rescue them. Historians and scientists believe Crocker Land was a superior mirage of a much more distant mountain range or of jagged Arctic ice (MacMillan 1933).

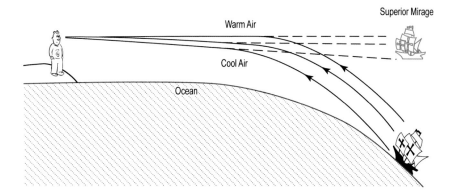

Fig. 9.14 Superior mirages form when warm air lies above cooler air. Light from distant objects can then refract downward so that normally unviewable objects become ethereally visible. If light passes through multiple layers of warm and cool air, even stranger effects are observed.

If you want to experience your own mirages (without having to brave the Arctic), find yourself in these types of locations: deserts, bays, or if you're feeling hearty, polar ice sheets. Generally mirages are most evident in places where large atmospheric temperature differences occur due to thermal heating of land or water. A pair of binoculars or a telephoto lens reveals mirages on the distant horizon often unseen by the naked eye. You can also see faint mirages if you place your eye close to a heated surface, such as a building or hood of a car, and look along the surface toward the horizon. In the shimmering of refracted air, a mirage of the Jones' new car may emerge.

9.11 Sundogs and Halos

On a day when high cirrus clouds cover the western sky and the light retreats in the late afternoon, look for three suns. These *sundogs*, also known as *parhelia*, lie 22° to the north and south of the actual Sun. Caused by sunlight refracting through flat crystals of ice, as shown in Fig. 9.15, the sundogs sometimes exhibit the spectrum of colors, especially if viewed through a pair of polarizing sunglasses. The hexagonal structure of the cirrus ice bends most of the sunlight 22° from the normal when the light strikes the edge of the flat crystals. Some of the light is refracted more than 22°, but none is refracted less. Like the rainbow discussed in Chapter 8, sundogs move with you. Only those ice crystals 22° (and a few more) from your line of sight to the sun will contribute to the effect; all the other refracted rays miss your eyes, unless you move to the right or left. On a

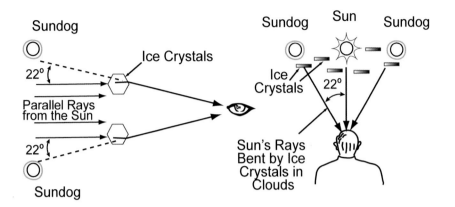

Fig. 9.15(a) Parallel light from the Sun is bent 22° by ice crystals that produce **(b)** images of the Sun 22° to the right and left from the true solar image. Look for sundogs in skies with cirrus clouds in the morning or late afternoon.

moonlit night, look for *paraselenae*, or *moon dogs*.

Another 22° refraction phenomenon occurs with long crystals of atmospheric ice rather than flat ice crystals. These also refract light (both sun- and moonlight), but in this case the crystals can be turned in any direction along two axes, as long as they lie horizontally. This orientation allows for a halo of light whose radius is 22°, centered around the sun or moon. The crystals also disperse the light, making the halo a composite of overlapping colored bands. When conditions are right, a rainbow seems to circle the sun. Look for halos on cold winter nights with winds blowing snow, or look for skies with high cirrus clouds.

A trip to the far northern or southern latitudes often rewards the traveler with other refractive displays such as subsuns, Circumzenithal arcs, and 44° halos. These can also (but more rarely) be seen in the temperate latitudes. Keeping an eye on the skies, however, will usually reward the faithful observer with wistful brushstrokes of light.

10 Polarized Light and Sunglasses

10.1 Sunglasses

We are awash in light, both visible and invisible. In its visible manifestation, light provides colors, shadows, and all the nuances of the visual experience. Yet in all of this visual life, we fail to note a hidden feature of light: it can be *polarized*. That is, as a wave of electromagnetic energy, its electric field component can be oriented in a specific direction but is always oriented perpendicular to the direction of the wave's travel. Figure 10.1 shows that white light can be blocked by two polarizer films held at right angles to each other.

Light emitted by most sources, such as the sun, light bulbs, and candles, is not polarized. The electric (and magnetic) field of the wave is oriented randomly in the plane perpendicular to the line of travel. Unpolarized light, however, becomes polarized by scattering from air molecules (blue sky), reflecting from

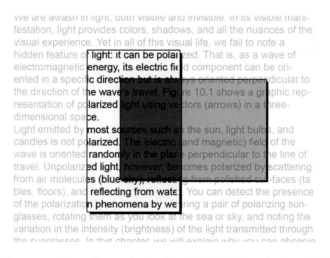

Fig. 10.1 A representation of two polarizer films held at right angles to each other. White light is blocked and is not transmitted in the regions where the two films overlap.

polished surfaces (tables, floors), and reflecting from water. You can detect the presence of the polarization phenomena by wearing a pair of polarizing sunglasses, rotating them as you look at the sea or sky, and noting the variation in the intensity or brightness of the light transmitted through the sunglasses. In this chapter, we will explain why you can observe this.

10.2 Polarized Light

E.T. Whittaker, in his introduction of Newton's *Optiks* (reprinted in 1979), says that the term "polarization" of light was taken from Newton's reference to magnetic poles: "… as the two *poles* of two magnets answer to one another" (italics added). Light of a single color can be described as an electromagnetic wave with a specified wavelength. The *wave* is the oscillation of electric and magnetic fields in time as they propagate through space, air or transparent objects. In previous chapters we've examined several features of the wave description of light: refraction, diffraction and interference. Another feature of this wave description of light is that it can be *polarized*, meaning the wave vibrations lie in one plane at right angles to the direction of propagation of light. Polarization of light is *transverse*, or perpendicular, to the light direction. In contrast, the vibrations in air of sound waves are *longitudinal*, that is, the waves vibrate along the direction

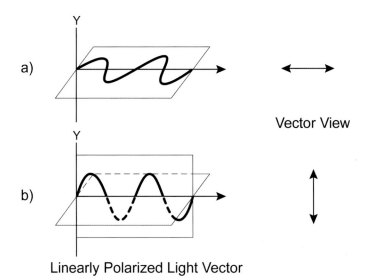

Fig. 10.2 Two examples of linearly polarized light where the electric vector of the light wave vibrates in one plane, **(a)** horizontal and **(b)** vertical.

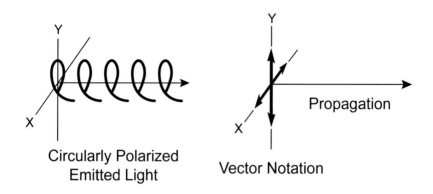

Fig. 10.3 The drawing on the left represents circular polarization, and the drawing on the right represents two-arrow vector notation.

of propagation, always pushing air forward.

Light can be *linearly* polarized, as shown in Fig. 10.2; the light wave oscillates on a plane in the same direction consistently. The figure shows a horizontal and a vertical plane depiction of two linearly polarized light waves. The arrows on the right side are conventional vector notations to indicate the orientation of the plane of polarization and the direction of the electric vector oscillation. When unpolarized light reflects off most objects, it becomes linearly polarized in the plane of the object. We will examine this phenomenon in the next section.

A *circularly* polarized beam of light, on the other hand, can be represented by two vector components of wave oscillations oriented at right angles to each other, as shown in Fig. 10.3. In this case, the direction of oscillation of the electric field is not consistently in one direction, but rotates in either the clockwise or counter-clockwise direction. If you could view the electric field arrow as the light came toward you directly, you would see the arrow as it oscillates 180° back and forth, rotating about its base like a spinning propeller.

10.3 Polarization by Reflection

If you live in a city or town, you may have noticed the intense reflection of sunlight off of wet streets and buildings and how the reflections can easily dazzle your vision. Light that reflects from a smooth, flat surface is often linearly (or partially) polarized; when there are many such reflections in our line of sight, we call it *glare*. Fig. 10.4 shows circularly polarized light incident on glass (it could also be unpolarized or partially polarized light). The reflected light is partially polarized in the plane of the reflecting surface. The refracted beam will

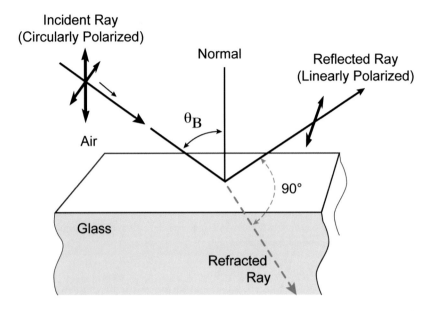

Fig. 10.4 Light incident on glass at the Brewster angle, θ_B, is reflected with complete linear polarization and refracted with a mixture of two perpendicular components.

contain a mixture of the two orientations. If the incident light is incident at the Brewster angle, θ_B, (named after Sir David Brewster), the reflected light is fully polarized, as shown in the figure. At angles other than the Brewster angle, the reflected light is partially polarized. But we often see light reflections near the Brewster angle, so their polarization is not uncommon.

The Brewster angle is related to the index of refraction, n, of the material (see Chapter 4). At the Brewster angle, the intensity of the reflected light goes to zero (i.e., the light disappears) when you look through a Polaroid film.

The reflection from water (or wet streets) can be directly tested by rotating Polaroid sunglasses. The dimming of the reflected light (and hence the presence of its polarization) is striking. Mountain climbers also use polarized *glacier glasses* to reduce or eliminate the glare of harsh sunlight from high-altitude snow and ice.

10.4 Polarization by Scattering

As we saw in Chapter 9, blue wavelengths of light from the sun are scattered preferentially in the atmosphere, more so than green or yellow or red light. This is why we see blue skies. This *Rayleigh* scattering from the air molecules also polarizes the scattered light. Is the sky-blue light polarized equally across the sky? No, it depends on the angular distance away from the sun that you're looking and if there are clouds or smog in the atmosphere. Let's examine the angular dependence first.

When an unpolarized light wave from the sun hits an atmospheric molecule (usually N_2), the wave is scattered. If the light scatters horizontally (parallel to the ground), the scattered wave will be polarized vertically. If it scatters vertically (down to the ground, or further up in the air), the scattered wave will be polarized horizontally. You can test this assertion using a simple Polaroid film (not the camera film) or sunglasses. Rotating the sunglasses or film while looking through it in the right direction of the sky will cause the sky's light to darken and then to brighten. What is the "right direction" in the sky? The most effective (full) polarization will be seen 90° away from the sun, since those waves will be fully polarized in the horizontal direction with no polarization in the vertical direction, the direction of light travel. Conversely, how will the light be polarized in the direction of the sky opposite the sun? (Test this at sunset or sunrise time.)

10.5 Polarization by Absorption

So far in this chapter we've discussed using Polaroid filters, film or sunglasses to determine if, and to what extent, light is polarized. What exactly do we mean by Polaroid, and how does it help us analyze polarized light?

By far the simplest way to observe polarized light is to use a piece of Polaroid, a material that absorbs light's electric field oscillations in one direction but allows the components oriented at right angles to pass through, as in Fig. 10.5.

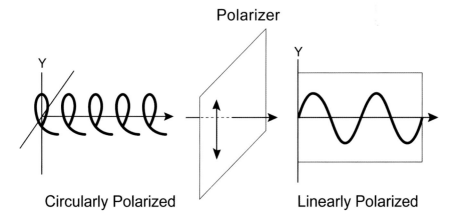

Fig.10.5 Circularly polarized light incident on a polarizer is transmitted as linearly polarized light.

Fig. 10.6 Two polarized films are opaque when oriented at right angles to each other. **(a)** Two aligned films, **(b)** partially twisted, and **(c)** at right angles.

In this way, the polarizer can transform circularly polarized light into linearly polarized light. These Polaroids are usually thin sheets of plastic containing long particles, rods or plates, aligned parallel to each other in a regular arrangement.

If two polarizers have their orientation axes aligned *parallel* to each other, then the first will absorb half the unpolarized light and transmit polarized light along their axes, and so the second polarizer will have no further effect, shown by Fig. 10.6(a). If the second polarizer is rotated by 90° so that the two are *crossed* with their axes perpendicular, then the second will totally absorb all the light transmitted by the first, as in Fig. 10.6(c). Each separately is transparent to light but when oriented at right angles to each other, they are opaque. The second polarizer is called an *analyzer*, because it analyzes the (polarized) light from the first polarizer. Consider the case of polarization by scattering: when you look at the blue sky through a Polaroid, which is the polarizer, and which is the analyzer?

Recalling the principles of polarization by reflection, the direct way to observe the polarization of diffusely scattered light is to look at the image of a light source reflected off a polished table (or the floor), as in Fig. 10.7. If the polarizer is rotated through 90°, the reflected image will dim. A view of the light source itself will show that it is unpolarized. Does the image of the light source through the polarizer dim as you rotate the polarizer?

10.6 Calcite and Double Refraction

If a beam of light is passed in a certain direction through a crystal of *Iceland Spar* (the optically clear form of the mineral *calcite* $CaCO_3$) we find that two beams of light emerge. One is undeviated while the other has been displaced from the original beam, as shown in Fig. 10.8.

The splitting of the beams can be seen by directing the beam of a light-emitting

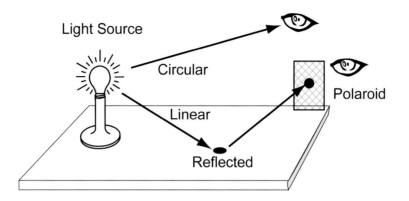

Fig. 10.7 Light source (circularly polarized) and reflected image (linearly polarized) viewed through Polaroid.

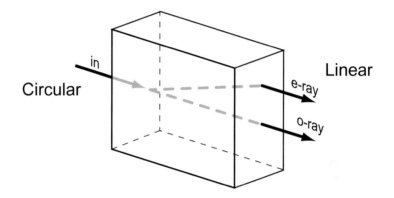

Fig. 10.8 A calcite crystal resolves a circular polarized beam of light into two linearly polarized beams.

diode through a calcite crystal or by simply holding the crystal over some text. The two-beam effect was originally seen by Erasmus Bartholine in 1669 and was described in 1690 by Christiaan Huygens in Treatise of Light. The phenomenon startled even Newton, who wrote, "If a piece of this crystalline stone be laid upon a Book, every Letter of the Book seen through it will appear double, by means of a double Refraction" (Republished in 1979).

A photograph of the displacement of a line seen through a calcite crystal is shown in Fig. 10.9. The phenomenon is easy to see and is demonstrated at most mineral and rock shows.

Fig. 10.9 Picture of a line taken through a calcite crystal. (Photo by Steve Beeson)

The double refraction phenomena served as a source of conflict between Newton, who favored the particle on corpuscular theory of light, and Huygens, who advocated the wave theory. Both theories are accurate, but Newton and Huygens made it a contentious period, as we saw in Chapter 4. In the modern era, the wave theory does a much better job explaining the phenomenon of double refraction.

10.7 Polarization and the Eye

Human vision does not distinguish between polarized and unpolarized light. However, materials such as Polaroids can absorb certain orientations of unpolarized and partially polarized light and allow us to *see* the resulting polarized light. Rotating a Polaroid or your polarizing sunglasses against a blue sky alternately dims and brightens the incoming light, providing a sense that one is viewing polarized sunlight.

Some people do claim to sense polarized light. This is said to present itself as a yellow light that they see when viewing the polarized sky. A phenomenon called *Haidinger's brush* arises because yellow pigment in the back of the eye selectively absorbs light with its electric field oriented in certain directions: the pigment polarizes by absorption. Since it's a yellow pigment, it absorbs *blue* light that's been polarized in a particular orientation. In some cases, people can see a yellow band when they look at a polarized, blue light source, such as the sky. This band may be vertical, horizontal, or diagonal, depending on the polarization of the blue sky! Look for it 90° from the Sun the next time you look up into a crystal clear sky.

11 Photons, Electrons, and the Atom

11.1 Light is Created and Destroyed

In earlier chapters, we mentioned the dual nature of light as the basis of a rivalry between seventeenth-century scientists Isaac Newton and Robert Hooke. They wondered about light. Is it a wave or a particle?[1] Can it be either? Can it be both? When is light most like a wave, and when is it most like a particle? Newton thought it was only a particle. Hooke preferred waves, as they explained much of the unexplained optical phenomena of the time. Other questions arose: How is light generated? How is it destroyed? (If it is.)

We delayed discussing the dual nature of light because, until this chapter, we were dealing mainly with the properties of light and their manifestations in the visible spectrum. For the remainder of this book, we will examine other wavelength regimes in the electromagnetic spectrum: the infrared, ultraviolet, and x-rays. And to understand how higher-energy light (such as ultraviolet and x-rays) is used to probe matter and its properties, it helps to understand light as particles, or packets of energy, rather than as waves. So in this chapter, we return to the questions of how light is created and destroyed, we pose new questions, and we examine how our understanding of the interaction of light and matter remains both a profound scientific advancement and a lingering mystery.

11.2 Packets of Energy

In either form — particle or wave — light is energy. Unlike other energies (that of motion or of potential motion), light is *radiative* energy, which means it is emitted by matter and absorbed by matter. While passing through space and not interacting with matter, it is an energy inert and undiminishing.[2] It is also

---

[1] Light can be described as either waves or particles.
[2] Except for its weak interaction with space-time, as described by Einstein's General Theory of Relativity.

without mass. *Photons*, which are what we call these packets of energy, have momentum, but no mass. And as we saw in Chapter 3, photons travel at the speed of light, which is 300,000 km/s, or 186,000 miles per second.

Their speed, however, has nothing to do with their energy. Unlike matter (e.g., electrons, protons, people), whose kinetic energy depends on their velocity, the speed of light and its energy are unrelated. In Chapter 4, we found that the energy of a light wave depends inversely on its wavelength: the longer the wavelength, the lower the energy. It is the same with light as energy packets. As Hertz discovered and Einstein explained, the energy of a discrete packet of light, or photon, is directly proportional to its frequency as a wave. These photons were useful in explaining how, for instance, a sheet of metal that was illuminated with light could emit electrons with specific energies regardless of how bright the light was.

In the classical idea of waves, this was nonsense. It was thought that, as the light got brighter, electrons with greater energy across a spectrum should come streaming off. Instead, it seemed the energy of the electrons emitted was directly related to the frequency of the light shined on the metal. It was kind of like ringing the bell with the sledgehammer at a county fair: instead of a stronger hit causing the puck to go higher, what mattered was the size of the hammer — and a smaller hammer seemed to work better! Back to the metal, if the light energy was contained in a localized quantum that gave all its energy to knocking out the electron from the metal, that would make more sense. It was Einstein's breakthrough in combining the classical wave theory of light with a new photon theory, or *quantum* theory, that began the twentieth-century revolution in the way we perceive and understand matter and energy.

11.3 The Electron

Both electrons and photons are at the heart of modern quantum theory, and both are described by wave and particle behavior. An electron, like a wave of light, can be made to diffract (by the nucleus of a large atom) and generate interference patterns. Its particle behavior is illustrated by the photoelectric effect, described in the previous section, whereby a photon, incident on a solid surface, ejects an electron from the material. Let us look at the properties of electrons before we delve into the interaction between electrons and photons, their ghostly brothers.

The properties of the electron have been studied for more than a century. Earlier scientists knew about electricity, but none tied the phenomenon to a single particle smaller than the atom. In one of the most serendipitous discoveries in science, a French physicist, Henri Becquerel, stumbled upon radioactivity when some *phosphorescent* rocks, wrapped in photographic paper and stowed in his desk drawer, emitted some kind of radiation while in the dark. Unwittingly, he had discovered how electrons (and other particles) can be spontaneously emitted from seemingly inert objects such as metallic rocks. The uproar caused by the discovery of radioactivity led to a surge of interest in the electron and the subsequent birth of modern physics.

One of the first pioneers in the study of electrons was J.J. Thomson, the British scientist who led Cambridge's Cavendish Laboratory, the pre-eminent physics laboratory of the time. In his 100-year-old work on the deflection of the electron by magnetic fields (Thompson 1897), he called the electrons *cathode rays.* Earlier (1891), the "electron was introduced as the name of the natural unit of electricity but the electron was not considered as a material particle." Later, American Robert Millikan showed that the electron has a fundamental unit of (negative) electrical charge and that its mass is 1/1845 that of the hydrogen atom. Much earlier, in the eighteenth and nineteenth centuries, scientists (namely Michael Faraday) had experimented with electric currents in the presence of magnets and developed two of the equations describing the electromagnetic nature of light. They found that electricity is easily deflected by magnetic fields, which anyone can see if you bring a magnet close to an old black-and-white TV screen (don't try it on a color screen, it may misalign the three cathode ray beams). James Clerk Maxwell (1831-1979) also found that an electric current is deflected by another nearby electric current; that is, in the presence of an electric field.

So it was when Ernest Rutherford, a student of Thomson's, quite literally punched a hole in the then-current *plum pudding* theory of the atom, which stated that negatively-charged electrons existed in a pudding of positive charge that made up the rest of the atom. In his famous experiment, Rutherford shot heavy alpha particles (helium nuclei) at atoms of gold, expecting them to go through the gold, slow down, or be slightly deflected. To everyone's surprise, he found that a handful of the alpha particles bounced straight back. His famous comparison was that "it was like firing a shell at a piece of tissue paper and watching the shell rebound." Rutherford had discovered the atomic nucleus and the vast emptiness of space between it and its orbiting electrons. There were plums (electrons and a positive nucleus), but there was no pudding.

Immediately, however, problems arose. As everyone knew then, the electron was negatively charged and deflected by an electric or magnetic field. If the nucleus was positively charged and the electron was negatively charged, eventually, the electron should either spiral into the nucleus (opposite charges attract), or fly off due to its own momentum. What kept the electron in an orbit around the nucleus? And how could light be emitted from such an atom, as in the photoelectric effect? A young Danish student of Rutherford's proposed a radical idea.

11.4 The Bohr Model of the Atom

The model of the hydrogen atom[3] developed in 1913 by Niels Bohr was that of an electron confined in a *potential well* caused by the positive proton in the nucleus. This potential well kept the electron from flying off into space. But what kept it from falling in? Bohr proposed that the electron could reside in only a few stable energy states dictated by a small number, called Planck's constant, which was like a quantum of orbit. Instead of the infinite number of energies most scientists believed Rutherford's orbiting electron could possess, Bohr's electron could only have finite energies and could reside only in fixed, stable orbits. Bohr had *quantized* the atom, just like Hertz and Einstein had quantized light. In fact, it was when Bohr remembered that gases emitted unique, discrete frequencies of light (spectra) that he made the connection between the quanta of light (photons) and the atom. When Bohr calculated the energy differences between these proposed electron orbits, they fit perfectly with the energy of the spectral lines emitted by hydrogen gas. Bohr forever tied the electron to the photon.

11.5 Light and Electrons

"There are three stages in the life of a light beam: it is created, it travels through space, and it is destroyed...light is created and destroyed only via its interaction with matter, from glowing gases in the sun to rhodopsin in the eye" (Sobel 1987).

Light's interaction with electrons is responsible for its creation and destruction. Yet the properties of the electron and photon are wildly different. The electron has charge and mass; the photon has neither. The photon has constant velocity in vacuum (the velocity of light, $c = 3 \times 10^8$m/s) independent of its energy; the kinetic energy of an electron is directly related to the square of its velocity. From tall radio towers beaming long radio waves to the x-rays streaming from a supernova explosion, light begins and ends with matter. But how does matter, something tangible and with mass, create light, something massless and ephemeral?

First, recall that light is a form of energy. Second, electrons in matter can be described as occupying energy levels — the *orbitals* of Bohr's hydrogen atom. Electrons can be raised to higher energies, (different orbits) by absorbing energy from a photon (destroying light). Electrons also can move to *lower* energies by emitting energy in the form of a photon (thus creating light) whose energy is

---

[3] The simplest atom is hydrogen, with one proton as the nucleus and one electron orbiting it.

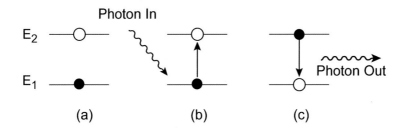

Fig. 11.1 Two electron energy levels in a solid. (a) An electron occupies the lower level and leaves the upper level empty (b) A photon is absorbed by the electron, which moves to the upper level (c) A photon is created by an electron moving from the upper to lower level with energy = E_2-E_1.

equal to the difference in energy between the two levels. Fig. 11.1 shows two electron levels separated in energy.

If a photon is incident on the electron in the atom, the photon gives up its energy (and vanishes) to the electron that moves to the upper level (Fig. 11.1(b)). In Fig. 11.1(c), an electron in the upper level moves to the lower level by giving up energy by emitting light. In this process, energy is conserved; the outgoing photon has exactly the same energy as the incoming photon. In other, more complex processes, the electron may move to intermediate energy levels and emit lower-energy photons (longer-wavelength light), but the total amount of energy in the photon and remaining in the electron levels is the same as the original incoming photon. Certain types of minerals and animals exhibit these processes of fluorescence or phosphorescence, i.e., the absorption of light and its re-emission at longer wavelengths.[4] We'll examine fluorescence in ultraviolet light in the next chapter.

11.6 Electrons and the Spectrum

How do we know that electrons reside in discrete energy levels, and that the energy of ejected photons is equal to the difference between those energy levels? First, Bohr's theory predicts the values of those energy levels for the hydrogen electron. Also, later quantum mechanical calculations predicted the electron energy levels for much more complicated atoms — those with many more electrons than hydrogen's one. If we could somehow measure the energy of photons emitted or scattered from an atom, we could compare the value to the predicted atomic energy level differences. As it turns out, Newton's prism took the first step in this process: a prism splits a beam of light into its constituent

---

[4] 'Phosphorescence' is fluorescence with a time delay.

wavelengths, ordered from red to violet for the longest to shortest wavelength, or in terms of energy, from lowest to highest.

Years after Bohr's theory emerged, scientists learned how to measure the energy of the colors split in a prism; the instrument became known as a *spectrometer*, a yardstick of light energy. As we saw in Chapter 7, spectrometers (some of which are made from diffraction gratings) display a spectrum in an ordered sequence, just as a prism does with sunlight. When analyzing light from a glowing gas or solid, however, the spectrometer displays sharp lines instead of a continuous rainbow. In the mid-nineteenth century, Gustav Kirchhoff and Robert Bunsen (for whom the Bunsen burner is named) studied the spectra of various elements as the elements burned or glowed and concluded that each has its own spectrum — the sharp lines seen in the spectrometer identified the element distinctly.

As early as 1814, astronomers were using spectrometers to deconstruct the light from distant stars and galaxies, and they found many different spectra in the heavenly bodies. Red stars tended to, of course, emit light in the red and orange part of the spectrum. Bright bluish-white stars emitted light across the entire spectrum, but especially in the blue. But each star's spectral signature was slightly different. What did this mean?

Stars are big, glowing, gaseous balls of hydrogen, helium, and other elements in trace amounts. The stars burn the gas in thermonuclear reactions, which convert the lighter elements like hydrogen and helium to heavier elements, such as carbon, sodium, and nitrogen. If the stars the nineteenth century astronomers observed with their spectrometers had different spectral signatures, that meant they must have been burning different gases in their interiors. Thus scientists found that the stars, including our sun, are composed of various elements in gaseous form, and each star has a slightly different composition — some with more helium, some with less carbon — and so their spectra differ. But scientists and astronomers still did not know *why* each element emitted a unique spectrum.

This, then, brings us back to Bohr. Recall that he theorized that, for hydrogen, we could imagine the sole electron existing in various energy states, or levels, as it orbited the nucleus. If the atom got bombarded with a photon or bumped by another atom, the electron might jump to a higher energy state, gaining the energy of the incoming photon or the kinetic energy of the intervening atom. Later, if it fell back to a lower energy, or even back to its original state, the atom would emit a photon whose energy was *exactly* equivalent to the difference between the high and low energy levels, E_2–E_1. With this theory, Bohr could predict the energy, or wavelength, of the photons that would be emitted by burning hydrogen gas. As we saw above, scientists had been doing just that nearly a century before Bohr's theory. All he had to do was match his predictions to what the observed spectrum of hydrogen told him were the energies (or wavelengths) of its emitted light. They matched nearly perfectly. Bohr had married the spectrum to the electron and presided over the transition between the classical ideas of light and the quantum view of nature.

11.7 X-ray Energies

Of course, the universe contains elements other than hydrogen. If we consider atoms that are more complicated than hydrogen and contain more electrons, which are negatively charged, we must remember that such atoms also contain nuclei of higher positive charge in order to keep the atom stable. The innermost electrons are held progressively more firmly as that nuclear charge increases. It takes larger increments of energy to move such electrons away from the nucleus into excited states. Conversely, larger quanta of energy are emitted when an electron drops closer to its ground state.

Figure 11.2 shows an energetic x-ray penetrating into the atom, colliding with an electron in the K-shell (the most strongly bound electron), and ejecting the electron. In this case, the photon is completely absorbed by the atom, which now has one less electron. For this scenario to occur, the incident x-ray photon energy must be greater than the *binding energy* of the K-shell electron, which is the energy needed to completely eject the electron from the atom. The atom loses energy when an electron makes a transition from the L3 occupied state to the K-shell vacancy, as shown in Fig. 11.2(b).

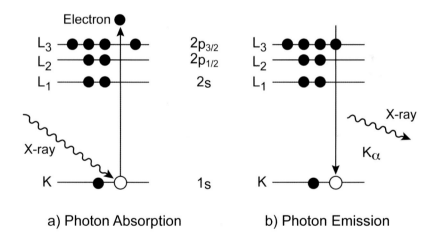

a) Photon Absorption   b) Photon Emission

Fig. 11.2(a) A photon is absorbed in transferring its energy to a K-shell electron **(b)** An x-ray (Ka x-ray) is emitted when an electron in the L3 shell transitions to the unfilled K-shell.

Fig. 11.3 Photo of the scanning electron microscope at Arizona State University.

Of course, when energetic x-rays or electrons are incident on an atom, many vacancies are created, and a number of x-rays are emitted. Like the spectra of visible light, the atom emits photons of unique and distinct energies, which appear as lines in an x-ray spectrum. These x-ray lines are called *characteristic x-rays*.

11.8 Characteristic X-rays

When an element is bombarded with high-energy electrons in an x-ray tube, it also emits characteristic x-rays with sharply defined energies. The characteristic peaks (termed *line spectra*) have specific energies that can be identified with each element, just as the nineteenth-century scientists found with visible light. Measurement of the line spectra energies allows one to identify the elements emitting the x-rays.

Most scanning electron microscopes (SEMs) have an x-ray spectrometer attached for the measurement and detection of x-rays. These instruments that measure x-ray energies contain liquid nitrogen (LN_2)-cooled detectors and can be easily identified by the cylindrical dewar (similar to a thermos bottle) about 8 inches in diameter and 12 inches tall. Figure 11.3 shows the SEM at Arizona State University with the operator sitting at the console. An LN_2 (liquid nitrogen) dewar is in the background.

12 X-rays, Ultraviolet Light, and Infrared

12.1 Beyond the Visible

What exactly happens when the dentist takes an x-ray of our teeth or a doctor examines a broken bone? What are x-rays? Can they harm us? How do they penetrate through bone and tissue?

What are black lights? How is infrared radiation used to analyze paintings? The answers are in the region of the electromagnetic spectrum that is beyond the visible.

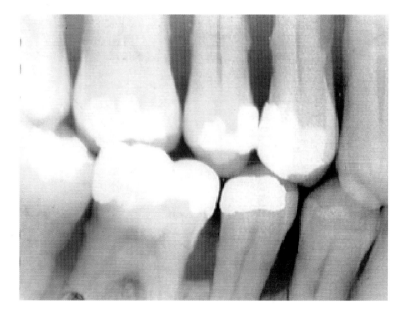

Fig. 12.1 A dental x-ray. The white areas are metal fillings in the teeth. Courtesy of R. Lorts, DMD

12.2 Discoveries beyond the Visible

Newton demonstrated in 1764 that white light could display a spectrum of colors. One hundred and forty years later, radiation was discovered beyond the visible region of the electromagnetic spectrum.

12.2.1 Infrared (IR) Radiation

The existence of radiation at wavelengths longer than the 700 nm of the red light we can see was discovered in 1800 by William Herschel. Herschel was using an experimental setup, shown in Fig. 12.2, to study the radiant heat produced in the visible spectrum.

Fig. 12.2 Sunlight incident on a prism is dispersed in the visible spectrum, which falls on a table containing thermometers that register the amount of incident radiant heat. (Herschel 1800)

Herschel found that the thermometers registered a strong temperature rise in regions beyond (at longer wavelengths than) the red portion of the visible spectrum (1800). In his words:

> It being now evident that there was a refraction of rays coming from the sun, though not fit for vision, were yet highly invested with a power of occasioning heat....To conclude, if we call light, those rays which illuminate objects and radiant heat, those which heat bodies, it may be inquired, whether light be essentially different from radiant heat? In answer to which I would suggest, that we are not allowed, by the rules of philosophizing, to admit two different causes to explain certain effects, if they may be accounted by one.

Herschel demonstrated the existence of the infrared — the invisible wavelengths that exist beyond the 700 nm wavelength of the red we can see. The energy of light is inversely related to wavelength so that the long wavelengths of infrared radiation correspond to energies lower than about one electron volt (1 eV).

12.2.2 Ultraviolet (UV) Radiation

Almost immediately after Herschel discovered infrared radiation beyond the red, J.W. Ritter in 1801 demonstrated that radiation existed at wavelengths shorter than the violet (wavelengths of 400 nm or energy greater than about 3 eV), a region now called ultraviolet.

Ritter found that silver oxide becomes dark much faster in the area where violet light falls on the sample. But he also found that the maximum of the intensity was far from where he could see any violet color.

Ritter had hypothesized a possible polarity in the spectrum based on radiation beyond the red end. He looked for invisible radiation beyond the violet end using paper soaked in silver chloride (essentially he used a photographic plate where the silver chloride would reduce to silver under illumination). Ritter found the paper blackened in the short wavelength region below the violet region (Ritter 1801).

12.2.3 X-radiation (X-rays)

Nearly 100 years after the IR and UV discoveries, W.C. Röntgen discovered x-rays (1896). He wrote, "A piece of sheet aluminum, 15mm thick, still allowed the x-rays (as I will call the rays, for the sake of brevity), to pass..."

Like the rays of visible light, x-rays follow straight lines, and obstacles in their paths cast shadows. In contrast to light, however, the absorption of x-rays depends only on the kind of atom in the absorber. The absorption of x-rays depends on the number of electrons (the atomic number, Z, of the atom). X-ray absorption is much easier to predict than light absorption, which depends on the chemical environment of the atom. For example, silicon absorbs visible light, and silicon dioxide is transparent to visible light.

In his 1896 article, Röntgen particularly noted that photographic dry plates are sensitive to x-rays. He included a photograph of the bones in the fingers of a living human hand (Fig. 12.3). The third finger has a ring on it. A modern x-radiograph (Fig. 12.4) shows a hand holding a telephone. You can see the bones in the arm as well as the hand. The metal components in the telephone are also clearly visible.

Our flesh is nearly transparent because it is composed of light elements, e.g., carbon and oxygen, which have low atomic numbers. Bones are absorbent because they contain calcium and metals (such as tooth fillings), which are opaque to x-rays. (See Fig. 12.1.) This property of x-rays was immediately used by medical professionals to examine fractures or to locate foreign bodies. They called it x-radiography or, more commonly, radiography, and within three months of Röntgen's discovery, x-rays were used as an aid to surgery (Brown 1975).

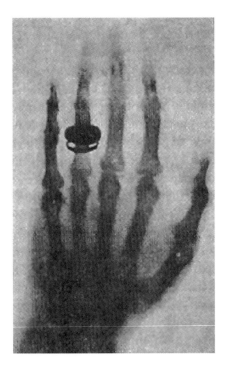

Fig. 12.3 X-ray photograph taken by Röntgen. The ring appears as a dark shadow because the material is highly absorbing.

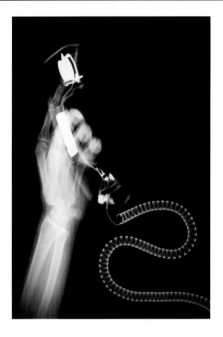

Fig. 12.4 An x-radiograph of a hand holding a telephone. (Photo by Steve Dunwell in Ideas 1, www.gettyimages.com/imagebank)

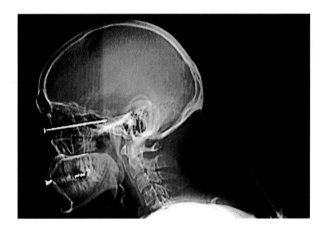

Fig. 12.5 X-radiograph showing the 3-inch nail that pierced the head of a Houston, Texas, carpenter. The nail missed the vital areas of the brain. (Aug. 2, 2001, New England Journal of Medicine)

A more recent demonstration of the use of x-rays in medicine is the radiograph of Fig. 12.5. The x-ray shows the nail that pierced the head but missed the eye itself. It also missed the carotid artery to the brain and the optic nerve.

X-rays are invisible. In his 1896 article, Röntgen says, "The retina of the eye is quite insensitive to these rays: the eye placed close to the apparatus sees nothing." This can be confirmed by closing your eyes during dental radiography — no flashes or glows are seen during x-ray exposure.

The discovery of x-rays completes the triad of phenomena outside the visible light spectra: infrared (IR) at low energies and long wavelengths next to the visible red; ultraviolet (UV) at high energies and short wavelengths, near violet; and finally, x-rays. Of course, the electromagnetic spectrum extends beyond these boundaries. For example, gamma rays are found at energies well beyond those of x-rays and have energies of a million electron volts (MeV) — a million times greater than visible radiation.

12.3 Light Absorption

The absorption of photons in solids depends on the number of electrons that can accept a transfer of energy from the photon. Not all electrons meet this criterion. Some electrons are so tightly bound to their orbital around the atomic nucleus that the photon energy is unable to break the bonds. Other electrons are involved in the bonding between atoms and cannot be freed by the energy of the incident photons. For example, ordinary window glass is transparent to visible photons but is strongly absorbent to ultraviolet radiation where the photons have energies only a few times greater than that of visible light. Thus window glass absorbs suntan-producing UV rays; you don't tan or burn when sitting in a car.

Photon absorption in the visible region of the spectrum depends on the atomic arrangement of the atoms and their bonding. Pure silicon (Si) is strongly absorbing, but silicon combined with oxygen (window glass) is transparent.

12.4 X-ray Absorption

In contrast to the description of the absorption of visible light, for energetic photons in the x-ray regime, photon absorption is much easier to predict and is independent of the details of atomic arrangement. Absorption of x-rays depends primarily on the electron concentration per unit volume. Since the concentration of atoms per unit volume only differ from each other by factors of 2 or 3, the electron concentration in materials can be estimated from the atomic number, Z, which gives the number of electrons per atom. For example, lead (Z=82) absorbs x-rays much more efficiently than aluminum (Z=13) and consequently is used in shielding around x-ray apparatus.

12.5 Fluorescence with Ultraviolet Light

The absorption of ultraviolet photons causes chemical reactions and, in some cases, leads to the emission of photons in the visible region of the spectrum. The absorption of light photons of *high* energy and the re-emission of photons of *lower* energy in the visible region is commonly called *fluorescence*.

Fluorescence occurs when light is absorbed at one energy, and part of the energy is re-emitted at longer wavelengths. When visible light is re-emitted, ultraviolet light has been absorbed. Fluorescence occurs at other energies; when it occurs for x-rays it is called x-ray fluorescence (XRF). In some cases, fluorescence occurs after the light source is removed. This slow emission of light is called *phosphorescence*. A common source of ultraviolet light is the black light source used in stores and discos. White shirts fluoresce, as do copy papers with whitener. Scorpions emit green light under black light UV illumination and are one of the few living species that fluoresce under ultraviolet light. Figure 12.6 shows a scorpion illuminated with ultraviolet light. Notice the vivid green color.

Fig. 12.6 Scorpion illuminated with ultraviolet light. (Photo by Misty Wing, Arizona State University)

12.6 Infrared Light

In the infrared (IR) portion of the spectrum, wavelengths are greater than 700 nm. These infrared photon energies, typically around 1 eV, are so low that the photons are not absorbed by most pigments used in paintings. The layers of paint are therefore relatively transparent to infrared radiation, which penetrates the upper layers but is absorbed by the dark colors of the charcoal preliminary drawings that reside beneath them. The remaining radiation is reflected by white or light-colored grounds. Because of the difference between absorbed and reflected radiation, the under drawing can be seen with the aid of infrared-sensitive photo equipment.

Figure 12.7 demonstrates the transparency of pigments to the infrared. Figure 12.7(a) shows an under drawing made using charcoal on a white ground. Figure 12.7(b) shows a layer of paint applied to the drawing, which is viewed with an infrared-sensitive video camera. The image of the under drawing, Fig. 12.7(c), can be seen clearly in the infrared display.

12.7 Infrared Space Exploration

The investigation of paintings with infrared reflectography covered a region in the infrared near wavelengths of 1000 nanometers or, more commonly, 1 micron (μm). The infrared portion of the spectrum extends to a hundredfold-longer wavelength (100 microns).

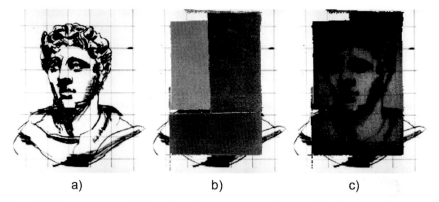

a)   b)   c)

Fig. 12.7 Demonstration of infrared reflectography. **(a)** Photograph of charcoal drawing in visible light. **(b)** Photograph of overlaid paint layers which hide the charcoal drawing. **(c)** Infrared reflectogram revealing the charcoal drawing beneath the paint layer. (Taken from Taft and Mayer, *Science of Paintings,*2000)

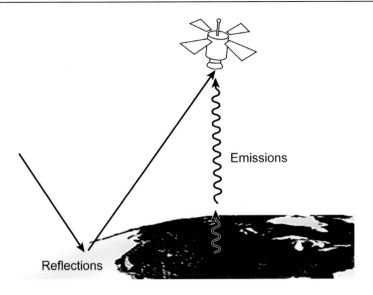

Fig. 12.8 Signals received by a satellite from reflection from the sun or thermal emission from various sources such as the surface of clouds.

Fig. 12.9 Thermal Emission Spectrometer (TES) images of water-ice cloud abundance on Mars taken by the Mars Global Surveyor.

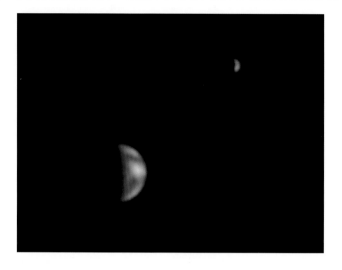

Fig. 12.10 An infrared image of the earth and moon taken on the Mars mission. (NASA Jet Propulsion Laboratory)

The Mars Global Surveyor was successfully launched into orbit around Mars on 11 September, 1997. A survey of the planet's surface and atmosphere was completed in January, 2001. As shown in Fig. 12.8, signals can be received from the surface because of reflection from the sun or thermal emission.

Figure 12.9 shows such thermal images of water-ice clouds on Mars from a Thermal Emission Spectrometer (TES) on the Mars Global Surveyor. The clouds in these images formed near the equator during the Martian northern spring. The thickest clouds appear over the large Martian volcanoes.

During the mission to Mars, a striking IR image was taken of earth and the moon (Fig. 12.10). It is a rare and unique view of our world and its planet.

We have shown some of the striking images that can be obtained outside the visible spectrum. In the next chapter we'll see what can be found with microscopes and x-ray detectors.

13 X-ray Emission: Earth, Moon, and Mars

13.1 The Moon

The composition of the Moon has been determined by placing a radioactive source on the surface and measuring the energetic particles and x-rays emitted from the soil. Figure 13.1 is a schematic of the sources and detectors that were used in 1967 to analyze the lunar soil. It was exciting because these surveyor measurements represented the first analysis of lunar soil. The results showed that the soils from the moon are similar to those from earth. These findings were confirmed by an analysis of moon rocks returned on the Apollo missions.

The moon's energetic particles come from a radioactive source that emits

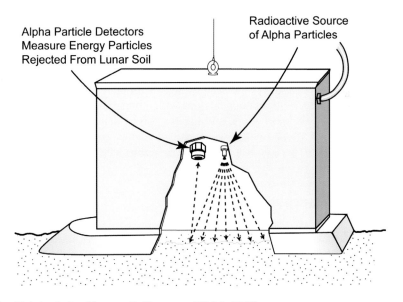

Fig. 13.1 Analysis of lunar soil. (Surveyor, NASA, 1967)

alpha particles with meV energies. On earth, we generate energetic particles in analytical instruments such as the scanning electron microscope (called an electron microprobe), and we measure the energies of x-rays emitted from the sample. An ion accelerator is used to accelerate a beam of helium ions to MeV energies and thus simulate the action of a radioactive source of alpha particles (doubly charged helium ions).

13.2 X-ray Emission

The general concept of particle-induced x-ray emission is shown in Fig. 13.2. Electrons are knocked out of occupied levels by energetic x-rays, electrons, or ions (either protons, H^+, or alpha particles, He^{++}). They then transition from filled to empty states and emit an x-ray. The energy of the emitted characteristic x-ray identifies the atom.

The incident particles must transfer to the atom an energy equal to or greater than the binding energy, E_B, of the electrons in the occupied levels. The electron energy levels are grouped in bands with each band differing from the others by about a factor of ten. The innermost band, which has the highest binding energy, is called the K-shell. The next level is the L-shell, then the M-, and so forth.

The x-ray energies are given by the binding energy difference between the levels. Figure 13.3 shows a schematic of the x-ray emission process for K-, L-, and M alpha transition x-rays. The beta transitions, K_β, L_β, and M_β, occur at slightly higher energies than alpha transitions because the full state has a lower binding energy. For example, the K_α transition is L-shell (full) to K-shell (empty), and the K_β transition is M-shell (full) to K-shell (empty).

The x-ray energies versus the atomic number, Z, are shown in Fig. 13.4. The energies of the emitted x-rays increase strongly with increasing atomic number because of the increase in binding energies with atomic number. There are also marked differences between the K and L x-ray for the same element. Each element has a unique set of emitted x-rays, called characteristic x-rays. Measurement of the x-ray energies allows identification of the element.

13.3 X-ray Fluorescence (XRF)

X-ray fluorescence equipment is used in art museums, for example, to examine the composition of pigments in paintings. Identification of pigments can help date a painting, thereby helping to determine if it's authentic. Figure 13.5 schematically shows the setup. An x-ray generator directs x-rays to the sample (a painting, in this case). These x-rays impinge on pigments in the painting, which emit characteristic x-rays that are analyzed in the x-ray detector.

Ion-, Electron- and X-ray-induced X-ray Analysis

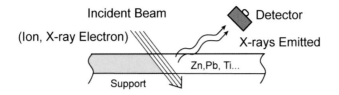

- Incident particle knocks electrons out of the occupied states around the atom leaving empty states (vacancies)

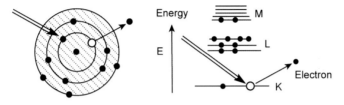

- Electron in occupied state makes transition to unfilled vacancy. X-ray is emitted to conserve energy.

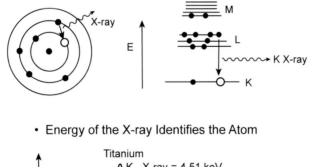

- Energy of the X-ray Identifies the Atom

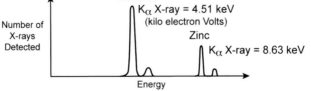

Fig. 13.2 Atom identification by ion-, electron-, and x-ray- induced x-ray emission.

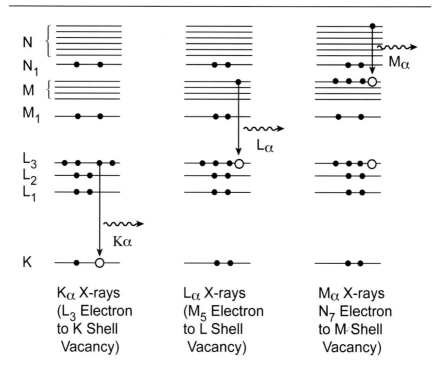

Fig. 13.3 Schematic of x-ray emission process where electrons in filled orbitals make transitions to inner shell orbitals with unoccupied (vacant) states. The x-ray energy equals the difference in binding energy between the filled outer shell and vacant inner shell.

13.4 Electron Microprobe and Electron Dispersive Spectrometry (EDS)

Chemical analysis (to determine the chemical makeup) of a specimen can be obtained by measuring the energy and the intensity distribution of the x-ray signal generated by a focused electron beam impinging the sample (Fig. 13.6). All the emitted x-rays with widely dispersed energies are detected and displayed as the number of x-rays per energy interval, hence the name Energy Dispersive Spectroscopy. With a different detector, called a Wavelength Dispersive Spectroscope, we can measure the wavelengths of the detected x-rays.

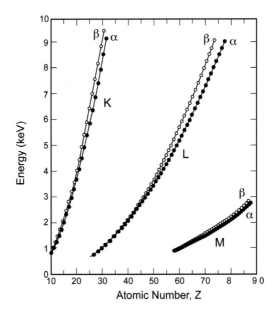

Fig. 13.4 Energies of the K_α, K_β, L_α, L_β, M_α and M_β x-ray lines as a function of atomic number. (Taken from Feldman & Mayer, Fundamentals of Surface and Thin Film Analysis)

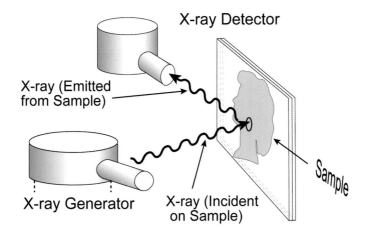

Fig. 13.5 Schematic of x-ray fluorescence.

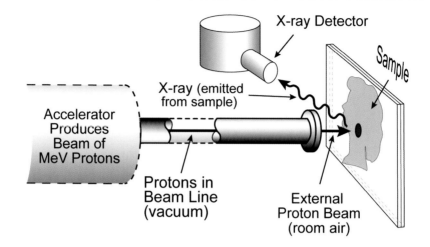

Fig. 13.6 Schematic of the setup for an external proton beam used in Particle-Induced X-ray Emission (PIXE) Remote Sensing.

13.5 Particle-Induced X-ray Emission (PIXE)

Several million electron volt (MeV) protons (singly charged hydrogen ions) or alpha particles (doubly charged helium ions which contain two protons and two neutrons) produced by ion accelerators are used to generate x-rays for materials analysis by Particle-Induced X-ray Emission (PIXE). There are hundreds of these accelerators in the U.S. and worldwide, located in universities and commercial laboratories. These accelerators produce a highly controlled beam of charged particles so that a known number of particles of known energy can be delivered to the sample. The analyses can be carried out in vacuum or in air. For in-air analysis, the ion beam — usually a stream of protons — passes through a thin membrane into air and onto the sample. Millions of electron-volt protons can pass through the thin membrane and several centimeters of air without substantial energy loss.

An *external beamline* is used to measure samples in air where energetic ions are taken from the vacuum of the ion accelerator beam and emitted into the air. Figure 13.6 shows the experimental setup that is used for large or vacuum-incompatible objects.

In Fig. 13.1 we saw an analysis of the moon by the surveyor alpha particles. Now in Figs. 13.7 and 13.8 we see the Alpha Proton X-ray Spectrometer on the Sojourner Rover on the Mars Pathfinder Mission.

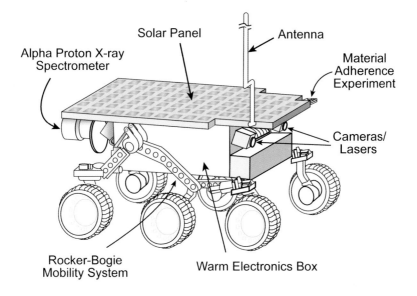

Solar Panel

Antenna

Alpha Proton X-ray
Spectrometer

Material
Adherence
Experiment

Cameras/
Lasers

Rocker-Bogie
Mobility System

Warm Electronics Box

Fig. 13.7 The Sojourner Rover measures about 65 cm x 48 cm and weighs about 10.6 kilograms on the surface of Mars. It is fitted with several different kinds of sensors for collecting data about the Martian environment. (NASA)

Fig. 13.8 The Mars Rover.

Fig. 13.8 shows the Mars Rover making an analysis of a Mars rock, nick-named "Yogi Bear." In the center of the figure, you can see the APXS (Alpha Proton X-ray Spectrometer) up against the rock.

The concept was the same on both missions, despite their 30-year separation: A radioactive source emits energetic alpha particles, which strike the lunar or Martian soil. These collisions with the soil cause emission of x-rays, protons, and other particles. The accumulation of data from detected particles allows the Sojourner's sensor to measure the geologic composition of rocks and surface soil.

Appendix A

Glossary

A

Aberration:
In microscopy, an effect that causes distortion or blurring of the image.

Aberration, Chromatic:
An effect created by the dispersion of the material from which the lens is made. This causes different wavelengths of light to be refracted to different degrees. The name chromatic comes from the colors evident with this type of aberration.

Aberration, Spherical:
An aberration caused by a lens being spherical. Spherical aberration is characterized by the edges of the image being magnified more than the center. Removed by using an aspheric lens or multiple spherical lenses.

Achromat doublet:
Two lenses of different refractive indices placed next to each other to eliminate chromatic aberration in a lens system. See *chromatic aberration.*

Accommodate:
In ophthalmology, to focus by contracting the crystalline lens of the eye.

Alpha, α:
Greek symbol used to denote an alpha particle.

Alpha Particle:
A doubly charged positive helium ion consisting of two protons and two neutrons. A term usually used in conjunction with radiation. Alpha particles can be ejected from certain radioactive isotopes.

Amplitude:
The measure of a wave from its lowest point to its highest. In physical mani-festations of waves, such as in water, the amplitude is a measure of distance. With light, the amplitude is a measure of the wave's energy.

Angstrom:
1×10^{-10} m. A convenient unit of measurement, as atom diameters are on the order of angstroms. Represented by the symbol Å.

Antisolar point:
The point 180° from the Sun, relative to an observer. The angular measure of rainbows can be made from the antisolar point, which corresponds to the shadow of one's head.

Anti-twilight arch:
The band of light above the horizon opposite a rising or setting sun. This phenomenon is due to an optical mix of Rayleigh-scattered blue wavelength light, and red wavelength light reflected from larger particles in the atmosphere. See *Rayleigh scattering*.

Aperture:
A small hole through which light passes. Apertures are used to form images or mask undesired light from an image.

Aqueous humor:
Liquid inside the cornea of the eye with a refractive index of 1.376.

Aragonite:
One of the two crystal forms of calcium carbonate, the other being calcite. Calcite is normally the stable phase. Both are present in the shells of certain sea creatures.

Atom:
The smallest particle of an element, which cannot be subdivided without cre-ating another element. Atoms have nuclei consisting of positively charged pro-tons and neutrons, which are not charged. The nucleus is surrounded by nega-tively charged electrons. Together, they make the atom neutrally charged. Almost all of the mass of an atom is in the nucleus.

Atomic Number:
The number of protons in an atomic nucleus. For a neutral atom, this also corresponds to the number of electrons orbiting the nucleus. For example, carbon, with symbol C on the periodic chart, has an atomic number of 6. This means that it has 6 protons, and the neutral carbon atom has 6 electrons.

Attenuation:
Weakening in force or intensity: "attenuation in the volume of sound."

B

Beta, β:
Greek symbol used to denote electrons.

Beta Particles:
High-energy electrons.

Bremsstrahlung Radiation:
An electric charge, as it is being accelerated or decelerated, produces radiation, which can be in the form of x-rays. This radiation is called Bremsstrahlung Radiation and usually creates an unwanted background in a spectrum.

C

Camera Obscura:
A room or box containing a hole in at least one of the walls that creates an image of the outside objects on the opposite wall.

Cantilever:
A projecting structure, such as a beam, that is supported at one end and carries a load at the other end or along its length.

Cathode-Ray Tube:
A vacuum tube in which cathode rays, in the form of a slender beam of electrons, are projected on a fluorescent screen and produce a luminous spot.

Charged:
Having acquired a net electric charge. Objects can become charged due to a transfer of electrons from one object to another. An example of this transfer is that which occurs while shuffling across a carpet or the rubbing of a plastic rod with a piece of fur.

Compound:
A substance consisting of atoms of different atomic numbers.

Compton Scattering:
The change in direction and energy of a photon due to the interaction with an electron.

Compound Microscope:
A microscope containing at least two lenses to form the image of a specimen. Invented by Robert Hooke in the mid-seventeenth century.

Concave:
An inward curvature. A concave lens or mirror has at least one surface that curves inward. A concave lens causes light rays to diverge, and a concave mirror causes light rays to converge. A mnemonic device to remember the difference between concave and convex is that caves are concave.

Cones:
The photosensitive cells in the eye's retina that respond to changes in the energy, or wavelength, of light. There are three types of cones: red-sensitive, blue-sensitive, and green-sensitive.

Converge:
To come together or come to a single place or point.

Convex:
An outward curvature. A convex mirror curves outward and a convex lens has at least one side that curves outward. A convex lens causes light rays to converge and a convex mirror causes light rays to diverge.

Cornea:
The outer layer of the eye. A major contributor to the refraction of light in the eye.

Coulomb:
A unit of electric charge. 6.25×10^{18} electrons will have a charge of one coulomb. A Coulomb per second flowing through a wire is defined as one Ampere.

Count:
The electronic pulse created in a detector by one quantum of radiation. Counts are summed to make spectral peaks.

Critical angle:

The angle beyond which light, traveling through an optically dense medium and incident on the interface with a less dense medium, will totally internally reflect. For light exiting glass to air, the critical angle is 41°. See *total internal reflection.*

Crystal:

A regular arrangement of atoms in three-dimensional space. Sometimes this regular arrangement gives the crystal a definite three-dimensional shape. For example, if the atoms arrange themselves in a cubic lattice, the crystal might be cubic.

Crystalline Lens:

The lens inside a human eye.

D

Depth of Focus:

Same as depth of field.

Depth of Field:

The spatial depth over which features can be considered in focus. The distance along the optic axis over which features are in focus.

Detector (Radiation):

Any device that detects incoming radiation and can create an electronic signal for every quantum of radiation, proportional to the energy of the quantum. Detectors can be of many types, gas and solid state. However, the detectors referred to here are all solid state, usually lithium drifted silicon.

Diffraction:

The bending of waves around corners. Sound can diffract around buildings and light can diffract around very small obstacles. The obstacle must be of a size similar to or smaller than the length of the wave involved. Diffraction can lead to interference phenomena.

Diffraction Grating:

A device which uses diffraction and interference phenomena to disperse a beam of light into its various wavelengths.

Diode:

A semiconductor device that emits light or can produce power when used as a solar cell.

Diopters:

A unit of measurement used in optometry. A diopter is the inverse of a lens' focal length, measured in meters.

Dispersion:

The change in refractive index with wavelength. Most refractive materials refract short wavelengths of light more than long wavelengths of light. This property is called the dispersion of the medium.

Diverge:

To spread or move apart.

Dopant:

A substance mixed in exceedingly small quantities with an otherwise pure substance. For example, phosphorus atoms are dopants in the semiconductor silicon.

E

Earth's Shadow:

The shadow of the Earth projected onto the atmosphere. Often seen for several minutes in a clear sky after sunset. The shadow is low on the horizon, bluish in color, and rises as the sun sets lower beyond the western horizon.

Eclipse:

A shadow or the act of blocking light. In astronomy, it refers to the phenomenon of the Moon blocking the Sun's light (solar eclipse), or the Earth blocking the Sun's light to the Moon (lunar eclipse).

Elastic Collision:

A collision between two bodies during which kinetic energy is conserved. During an elastic collision the bodies do not stay together.

Electromagnetic Radiation:

The type of radiation we know as light: x-rays, radio waves, ultraviolet and infrared, among others. Electromagnetic radiation consists of alternating electronic and magnetic fields at right angles to each other. Both of these fields consist of waves and impart electromagnetic radiation with a specific wavelength and frequency. Electromagnetic radiation travels at 3×10^8 meters per second in a vacuum. See *electromagnetic spectrum*.

Electromagnetic Spectrum:

The entire range of wavelengths through which electromagnetic radiation can be manifested. The longest wavelengths are the lowest energies and the shortest wavelengths are the highest. Starting at the longest wavelengths, there are radio waves, which proceed through the microwave region. The next highest energy radiation is infrared followed by visible light, then ultraviolet radiation, x-rays and then gamma rays.

Electron:

An elementary particle having a rest mass of 9.11×10^{-31} kilograms and an electric charge of -1.602×10^{-19} coulombs. Electrons, being negatively charged, cancel out the positive charge of an atoms nucleus. In the structure of an atom, electrons surround the positively charged nucleus.

Electron, Auger:

An electron that has been ejected from an atom by the Auger process. In this process, an inner shell electron is emitted from an atom. An outer shell electron then fills the inner shell vacancy, and in the process, the extra energy causes another outer shell electron to be ejected. This outer shell electron is referred to as the Auger electron and the energy of this electron is measured to determine elemental composition.

Electron, Backscattered:

An electron that has been scattered directly from a nucleus. Backscattered electrons lose very little energy as a result of the backscattering process and are usually used for compositional or Z contrast imaging in electron microscopy.

Electron, Secondary:

An electron that has been ejected from a sample as a result of bombardment by an incident beam, They are ejected only after considerable scattering and energy loss inside the material. Secondary electrons are used for topographic imaging in electron microscopy.

Electron Volt:

A unit of energy. An electron volt is equivalent to the energy required move an electron through a potential difference of one volt. An electron volt is equal to 1.602×10^{-19} joules.

Element:

A substance consisting of atoms all of the same atomic number. Elements can combine to make compounds.

Energy:

A physical quantity defined as the ability to do work. Some of the types of energy include heat (the motion of atoms), kinetic (the movement of a mass through space), potential (a mass with the capability to move in response to a force), electrical, chemical, and others. Energy can be transformed from one form to another but not destroyed. Light is defined as electromagnetic energy.

Extended source:
A light source that emits from a finite area or volume. The Sun is an extended light source as seen in the sky. See *point source*.

F

First surface:
The surface of a lens or mirror that first encounters incident light. A household mirror has a first surface of glass over a layer of silver.

Fluorescence:
An ability that some transparent bodies have to modify incident light to produce light lower in energy and different in color from the mass of the material. For example, the hyaline layer is a glassy translucent material found in scorpions that causes fluorescence under ultraviolet light.

Focal Length:
The distance from a lens at which parallel rays converge after they pass through the lens or reflect from a mirror.

Focal Plane:
The plane at which a lens forms images. A movie screen is at the focal plane of a projector's lens.

Focal Point:
A point of convergence of parallel rays of light (or other radiation) or a point from which it diverges. The location at which rays parallel to the optical axis of an ideal mirror or lens converge to a point.

Frequency:
The number of waves passing a point in time. For light, this is usually expressed in Hertz (Hz), the number of waves per second. Related to wavelength by the equation $f = v/\lambda$ where f is the frequency, v is the wave velocity, and λ is the wavelength.

Fresnel Lens:
A lens in which only the outer layer of glass or plastic is used to refract light. Used in lighthouses, lanterns, and as flat lenses.

G

Gamma, γ:
Greek symbol used to denote high-energy photons.

Gamma Radiation:
High-energy electromagnetic radiation emitted by excited atomic nuclei.

Gauss:
A unit of magnetic field strength.

Grazing Incidence:
An object or light striking a surface at a highly oblique angle. Light encountering a surface at grazing incidence is nearly 100% reflected.

H

Hole:
A vacancy in the electron population of a semiconductor, which acts as a positively charged current carrier.

Hooke's Law:
An explanation of how springs behave. By applying force (F) to the spring, the spring will stretch by length (x). When you double the force (F) you double the length (x). Hooke's law applies to both stretchable springs and compressible springs. The only limit is when the spring is stretched beyond its elasticity and becomes deformed.

Hyperopia:
A malady of the eye in which the focal plane rests behind the retina. Commonly known as "farsightedness." Corrected with convex lenses.

I

Incandescence:
Giving off light due to the attainment of an extremely high temperature.

Incandescent:

See *incandescence*.

Incident:

Describes electromagnetic energy or ions which are directed at and strike a target or sample. As opposed to scattered or reflected energy or particles, which are leaving the target or sample. Incident photons or other particles can be scattered, reflected, or absorbed.

Index of Refraction:

See *refractive index*.

Inelastic Collision:

A collision between two bodies in which kinetic energy is not conserved, but momentum is.

Infrared:

Radiation in the electromagnetic spectrum between visible light and radio waves.

Interface:

The boundary between two different materials or media.

Interference:

The constructive and destructive addition of waves. Interference can lead to patterns sometimes called diffraction patterns, as in Young's double slit experiment.

Ion:

In general, any charged particle. In aqueous chemistry, usually a soluble charged particle. In physics, usually an atom with an electron added or removed. If an electron is added the ion is negatively charged; if an electron is removed, the ion is positively charged. When this happens, the atom is said to be ionized.

Ionization:

The creation of ions. In physics, the gain or loss of an electron by an atom. In chemistry, the gain or loss of an electron by a group of atoms. Gaining an electron creates a negative ion and losing an electron creates a positive ion.

Iridescence:

The property of producing colors due to the interference effects of light reflecting off of different layers at different angles. See *interference*.

Iris:
The network of muscle tissue in the eye that opens and closes the pupil and is responsible for the color of one's eyes. See *pupil*.

Irradiate:
To expose to any type of radiation.

Isotope:
A different form of the same element. Isotopes cannot differ in the number of protons or electrons they possess, but they do differ in the number of neutrons in the nucleus. See *radioisotope*.

J

Joule:
A unit of energy. The amount of energy it takes to exert a force of one Newton through a distance of one meter. Also equal to one Watt of electrical energy expended for one second. See *Newton*.

K

Keratin:
A protein which makes up most of the structure of hair and fingernails.

L

Lambda, λ:
Greek symbol used to denote wavelength.

Laser:
A device that utilizes the stimulated emission of light to create and amplify a beam of light. The laser-producing material is often gallium arsenide, helium-neon, or ruby.

Lattice:
An imaginary regular arrangement of points in three-dimensional space. Most solids can be characterized by their lattice structures.

Lattice Point:
An imaginary point in three-dimensional space that defines a structure point of a lattice. Any lattice point is equal to any other lattice point. It will have the same number of lattice points around it in the same directions the same distance away as any other lattice point.

LED:
Light emitting diode. A semiconductor device that emits light of a specific energy when electrical current flows through it.

Lepton:
Any of a class of elementary particles that consist of the electron, neutrino, muon, and tau particle.

M

Magnification:
The degree to which an image is larger than the object. Calculated as (image length) / (object length).

Meter:
A unit of length defined in 1800 as one ten-millionth of the meridian distance from the equator to the North Pole. One meter, symbol m, is approximately 39 inches in length.

Meteor:
The streak of light in the sky resulting from the incandescent particles given off by an incoming piece of space debris. Before it strikes the ground, the debris itself is called a meteoroid. After it has struck the ground, it is called a meteorite.

Meteorite:
An object from space which has entered the Earths atmosphere and struck the ground. Meteorites can be "finds" or "falls". Finds are found some unknown time after striking the Earth, and are not seen to strike the ground. Falls are seen to fall to the Earth and then are recovered where they were seen to fall.

Micro:
The Greek symbol μ, which denotes 10^{-6}.

Microbiology:
The study of microscopic organisms and single cells.

Micron:
1×10^{-6} m. Symbol μm.

Microscope:
An arrangement of lenses deigned to view objects at high magnification. A simple microscope contains only one lens. A compound microscope contains more than one lens. Electron microscopes contain electron lenses. By extension,

any device designed to view objects at larger magnification, even without lenses. (i.e., Scanning Tunneling Microscope, STM, Near Field Scanning Optical Microscope, NSOM.)

Mirage:
An image distorted by light refracted through heated or cooled air. *Inferior* mirages are formed in layers of cool air over warm air and are often seen in the desert. *Superior* mirages are formed in layers of warm air over cool air and are often seen over the ocean or frozen landscapes.

Molecule:
The smallest particle of a substance that retains the chemical and physical properties of the substance and is composed of two or more atoms; a group of like or different atoms held together by chemical forces.

Momentum:
That property which is conserved during elastic or inelastic collisions. It is defined as the product of the mass and the velocity of the body, and is represented by p. $p = m*v$. Also equal to a force over a unit time. $P = f*\Delta t$. In quantum mechanics, $p = h/\lambda$ where h is Planck's constant and λ is wavelength.

Mylar:
A trademark used for a thin, strong polyester film.

Myopia:
A malady of the eye in which the focal plane rests in front of the retina. Commonly known as "nearsightedness." Corrected with concave lenses.

N

Nanometer:
1×10^{-9} meters. Symbol nm.

Neutron:
An elementary particle having mass 1.675×10^{-27} kg, slightly greater than that of the proton, and no charge. The nuclei of atoms contain protons and neutrons.

Newton:
A unit of force in the M.K.S. system. The force required to impart to a mass of one kilogram an acceleration of one meter per second squared.

Normal:

Perpendicular, or at a right angle. Lines can be normal or perpendicular, planes can be normal, and lines can be normal to planes.

Nucleus:

The very small positively charged core of an atom. The atomic nucleus consists of positively charged protons and uncharged neutrons. Most of the mass of an atom is in the nucleus.

O

Ommatidium:

A single unit in the compound eye of an insect. It consists of a facet, a lens and a photoreceptor.

Opaque:

The property of blocking light. An opaque object passes no light through it.

P

Penumbra:

The part of a shadow where some of the light from an extended light source is blocked. In astronomy, the penumbra is the area on the ground where observers see a partial solar eclipse. See *umbra, eclipse*.

Period:

The time to complete one cycle of motion. The period of the Earth's rotation is 24 hours.

Phosphor:

A material that emits lower energy light when illuminated by high energy light. The insides of fluorescent light bulbs are coated with phosphor.

Photon:

A quantum of electromagnetic radiation. Photons are characteristic of all electromagnetic radiation, from the longest radio waves up to the shortest gamma rays. The shorter the wavelength of the radiation, the more energetic the photon is.

Piezoelectric effect:

Certain crystals with asymmetric unit cells have this property. When a compressive stress is applied to these crystals, it effectively shifts the positive ions in one direction and the negative ions in the other. This produces a temporary charge buildup on the opposite surfaces of the crystal. A tensile stress reverses the charge. The corollary effect can occur; an applied potential difference can change the shape of the crystal, or apply a force against a restraint.

Piezoelectricity:

Electricity or electric polarity due to pressure especially in a crystalline substance (as quartz).

Pigment:

A substance used as coloring. Dry coloring matter, usually an insoluble powder, to be mixed with water, oil, or another base to produce paint and similar products.

Pinhole Camera:

A device used to record or view images of light caused by a pinhole that forces light rays to cross.

PIXE:

Particle Induced X-ray Emission. The process whereby an incident ion causes an inner shell electron to be ejected from an atom. The resulting electron transition causes a characteristic x-ray, which is analyzed.

Planck's Constant:

A constant of proportionality between the wavelength of electromagnetic radiation and the energy by the equation $E = h*c/\lambda$. Where E is the energy of the photon of electromagnetic radiation, λ is the wavelength, c is the velocity of light and h is Planck's constant, 6.626×10^{-34} Joule-seconds.

Point Source:

An ideal light source that emits from an infinitesimally small area. Stars as seen in the sky are essentially point sources. See *extended source*.

Polarize:

To cause radiation to be polarized. If light passes through a polarizing filter, the alternating electric and magnetic fields will no longer be oriented in all directions perpendicular to the direction of motion, but will be oriented more strongly in specific directions. Light can also be polarized by passage through the atmosphere and by reflecting off of surfaces at a very shallow angle. See *electromagnetic radiation*.

Positron:

Positively charged particles with the same mass as an electron.

Presbyopia:

A malady of the eye in which the crystalline lens loses the ability to focus properly. Occurs commonly with age. Corrected with convex lenses.

Primary rainbow:

The rainbow found 40–42° from the *antisolar point.* Usually the only rainbow in the sky. Sometimes seen with the *secondary rainbow.*

Prism:

A device which uses the dispersion property of glass to split light into its component wavelengths.

Proton:

An elementary particle of mass 1.673×10^{-27} kg or about 1,836 times that of an electron. The proton has a charge equal in magnitude as that of an electron only it is positive instead of negative, being $+1.602 \times 10^{-19}$ Coulombs. The nuclei of atoms contain protons and neutrons.

Probe:

A usually small object used especially for exploration to obtain specific information. Definition created in 1580.

Pupil:

The aperture of the eye. See *iris.*

Q

Quantum:

A discrete unit that exists at very small dimensions. A quantum of electromagnetic energy (light) is the photon.

R

Radiation:

The emission of particles from a source. The particles may have mass, as in the case of neutrons or alpha particles, or they may be massless as in the case of photons such as x-rays and gamma rays.

Radioisotope:

A different form of a radioactive element that emits radiation. The radioisotope differs in the number of neutrons in the nucleus of an atom.

Raster:
A scan pattern (as of the electron beam in a cathode-ray tube) in which an area is scanned from side to side in line from top to bottom; also: a pattern of closely spaced rows of dots that form the image on cathode-ray tube (as of a television or computer display).

Rayleigh Scattering:
The scattering of light or electrons due to interactions with particles whose size roughly corresponds to the wavelength of the photons or electrons. The amount of scattering follows the inverse of the wavelength to the fourth power. Blue sky, smoke, and some animal coloring are due to Rayleigh scattering.

Reflection:
The act of striking a surface and being repelled from it. When an incident photon leaves a surface on the same side (as opposed to the opposite side, being transmitted, or being absorbed in the material) it has been reflected. Different surfaces reflect light to different degrees and different surfaces reflect different wavelengths of light. For example, if a surface reflects light of 4,500 Angstroms, we say it reflects blue light.

Refraction:
The act of passing through a transparent material and changing direction at the interface. The degree to which light refracts at an interface is determined by Snell's law.

Refractive Index:
A numerical measure of refraction, related to the velocity of light in a material. Denoted by the letter n. $n = v_0/v_n$, where v_0 is the velocity of light in a vacuum, and v_n is the velocity of light in the medium of refractive index n. n in a vacuum is 1, while n in water is 1.33. See *refraction*.

Resolve:
The ability to discern two separate, closely spaced objects. Related to resolving power, the ability of a device to optically resolve distinct objects.

Retina:
The layer of photosensitive cells at the rear of the human eye.

Rods:
The photosensitive cells in the eye's retina that respond to changes in the brightness of light.

S

Secondary Rainbow:

The rainbow seen outside a *primary rainbow*. This rainbow is 49-51° from the *antisolar point*, and its colors are reversed relative to the primary.

Selection Rules:

In x-ray emission from atoms, these rules determine which electron transitions between energy states are allowed.

SEM-EDS:

Scanning Electron Microscopy-Energy Dispersive Spectroscopy. The process whereby an incident electron causes an inner shell electron to be ejected from an atom. The resulting electron transition causes a characteristic x-ray, which is analyzed. This takes place in an electron microscope as the electrons used for imaging also cause the emission of x-rays.

Semiconductor:

A material that can be an insulator in very pure form, but is usually doped with very small amounts of impurities in order to cause it to conduct with more or less resistance depending on the amount of the dopant. Silicon and germanium are two elemental semiconductors. Unlike a metal, the resistance of a semiconductor decreases with an increase in temperature.

Solar Cell:

A semiconductor device used to produce power when irradiated by sunlight.

Sound:

Vibrations of air molecules that we hear.

Spectrometer:

An instrument used for measuring the energies and wavelengths of photons.

Spectroscope:

An optical instrument that produces a spectrum of the wavelengths of light.

Spectrum:

Plural, Spectra. Any visible manifestation of radiation, which is arranged according to wavelength or energy. Can refer to the visible spread of radiation caused by a prism or diffraction grating in visible light, or to the graph created by any method of plotting radiation intensity versus photon energy or wavelength.

Speed of Light:

The speed at which electromagnetic energy travels in a vacuum, which is 3 x 10^8 m/s. Nothing that carries information can travel faster than the speed of light.

Sundog:

An atmospheric optical phenomenon that mimics the image of the Sun. Caused by light refracting through horizontal ice crystals in the atmosphere, which often disperse the light as well. Also known as *parhelia*, sundogs are found 22° from the solar disk in skies with cirrus clouds.

T

Total Internal Reflection:

A phenomenon that occurs when light is incident on an interface between two media of different refractive indices. If the light is traveling through a denser medium and strikes the surface at an angle greater than the critical angle, the light will be totally reflected inside the medium. Optical fibers take advantage of this property to propagate light through the fiber.

Translucent:

The property of passing light with some loss of the light. One can normally see part way into translucent objects, but not all the way through.

Transparent:

The property of passing light.

U

Ultraviolet:

A region of the electromagnetic spectrum in which the photons have higher energy than the visible region but lower energy than x-rays. Opinions vary on exactly where the ultraviolet region begins and ends, but the wavelengths are about 100 angstroms to about 4000 angstroms.

Umbra:

The central part of a shadow where all light from an extended (or point) source is blocked. See *penumbra, eclipse*

Unimpeded:

Not slowed or prevented; a good vacuum is necessary to allow the electrons to move unimpeded down the column of a scanning electron microscope.

V

Vacuum:
A condition of rarified gas. A perfect vacuum can never be attained, as every surface has water and various other substances on it in minute quantities, which will evaporate under conditions of vacuum.

Visual Ray:
Theorized by Empedocles in the fifth century BC to explain the visual perception. An human eye would emit a visual ray to intercept a light ray from an object; the returning ray combination would inform the observer of the perceived object.

W

Wave Number:
The number of waves in a given length. Equal to the reciprocal of the wavelength, $1/\lambda$.

Wavelength:
The distance between two successive peaks or troughs on a wave. Related to frequency by the equation $f = v/\lambda$, where f is the frequency, v is the velocity of the wave and λ is the wavelength.

X

X-ray:
Any quantum of electromagnetic radiation between the wavelengths of 1×10^{-9} m and 1×10^{-11} m or between the energies of approximately 1 keV and approximately 100 keV.

X-ray Diffraction:
The reflection of x-rays from planes of atoms.

XRF:
X-Ray Fluorescence. The process whereby an incident photon causes an inner-shell electron to be ejected from an atom. The resulting electron transition causes a characteristic x-ray, which is a signature for and identifies the atom.

References

Boynton R (1992) Human color vision. Optical Society of America, Washington DC

Boys CV (1959) Soap bubbles. Dover, New York

Brill TB (1980) Light: Its interaction with art and antiquities. Plenum Press, New York

Brown JG (1975) X-rays and their applications. Plenum/Rosetta Edition, New York

Chaisson E (1998) The Hubble wars. Harvard University Press, Harvard

Crease RP, Mann CC (1986) The second creation. Macmillan, New York

Dawkins R (1998) Unweaving the rainbow. Houghton Mifflin Company, Boston.

Falk DS, Brill DR, Stork DG (1986) Seeing the light. John Wiley & Sons, New York.

Feldman LC, Mayer JW (1986) Fundamentals of surface and thin film analysis. North-Holland, New York

Goldberg B (1985) The mirror and man. University of Virginia Press, Charlottesville.

Gowing, L (1952) Vermeer. Faber, London.

Hecht E (1987) Optics. Addison-Wesley, Reading, Massachusetts

Herschell W (1800) Experiments on the refrangibility of the invisible rays of the Sun. Phil. Trans. Roy. Soc., London 90:284

Herzberger M (1966) Optics from Euclid to Huygens. Applied Optics 5, 9:1383–1393

Huygens C (1690) Treatise of Light. University of Chicago Press, Chicago

Isenberg C (1992) The science of soap films and soap bubbles. Dover Publications, New York

Johansson SAE, Campbell JL, Malmqvist KG (eds) (1995) Particle-induced x-ray emission spectrometry (PIXE). John Wiley and Sons, Inc., New York

Kirsh A, Levenson RS (2000) Seeing through paintings. Yale University Press, New Haven

Lee jr RL, Fraser AB (2001) The rainbow bridge. Pennsylvania State Press, University Park

Lynch DK, Livingston D (1995) Color and light in nature. Cambridge University Press, Cambridge

MacMillan D (1933) Four years in the white north. Hale, Cushman, and Flint, Boston

Meinel A, Meinel M (1983) Sunsets, Twilights, and Evening Skies. Cambridge University Press, Cambridge

Micscape. Microscopy UK Magazine. http://www.microscopy-uk.org.uk

Millikan RA (1911) Phys. Rev. 32:349

Minnaert M, Seymour L (1993) Light and color in the outdoors. Springer-Verlag, New York

Minnaert M (1954) The nature of light and color in the open air. Dover, New York, 1954

Nassau K (1983) The physics and chemistry of color. John Wiley & Sons, New York

Newton I (1952) Opticks. Dover, New York

Park D (1992) The fire within the eye. Princeton University Press, Princeton.

Parvianinen P. Mirages in Finland. Virtual Finland. http:// virtual.finland.fi/fin-fo/english/mirage.html

Pedrotti FL, Pedrotti L (1993) Introduction to optics. Prentice Hall, New Jersey.

Perkowitz S (1996) Empire of light. Henry Holt and Company, New York

Ritter, JW (1801) Naturforschende Gesellschaft Zu Jena.

Roche S, Courage G, Devinoy, P (1985) Mirrors. Rizzoli, New York.

Rontgen WC (1896) On a new kind of rays. Nature. 53:274

Rossing TD, Chiaverina, CJ (1999) Light science: Science and visual arts. Springer-Verlag, New York.

Rossotti H (1983) Colour. Princeton University Press, Princeton

Sabra AI (1981) Theories of light: From Descartes to Newton. Cambridge University Press, Cambridge

Simon H (1971) The splendor of iridescence. Dodd, Mead and Company, New York

Sobel D (1995) Longitude. Walker & Co., New York

Sobel MI (1987) Light. University of Chicago Press, Chicago

Spiro IJ, Schlessinger M (1989) Infrared technology fundamentals. Marcel Dekker, Inc., New York

Steadman P (2001) Vermeer's camera. Oxford University Press, Oxford.

Taft WS, Mayer JW (2000) The science of paintings. Springer-Verlag, New York

Thompson JJ (1897) Philosophical Magazine 44:293

Index

Printed in China